AGRICULTURAL BIOTECHNOLOGY

Strategies for National Competitiveness

Committee on a National Strategy for
Biotechnology in Agriculture

Board on Agriculture

National Research Council

NATIONAL ACADEMY PRESS
Washington, D.C. 1987

NATIONAL ACADEMY PRESS ● 2101 Constitution Ave., NW ● Washington, DC 20418

NOTICE: The project that is the subject of this report was approved by the Governing Board of the National Research Council, whose members are drawn from the councils of the National Academy of Sciences, the National Academy of Engineering, and the Institute of Medicine. The members of the committee responsible for the report were chosen for their special competences and with regard for appropriate balance.

This report has been reviewed by a group other than the authors according to procedures approved by a Report Review Committee consisting of members of the National Academy of Sciences, the National Academy of Engineering, and the Institute of Medicine.

The National Academy of Sciences is a private, nonprofit, self-perpetuating society of distinguished scholars engaged in scientific and engineering research, dedicated to the furtherance of science and technology and to their use for the general welfare. Upon the authority of the charter granted to it by the Congress in 1863, the Academy has a mandate that requires it to advise the federal government on scientific and technical matters. Dr. Frank Press is president of the National Academy of Sciences.

The National Academy of Engineering was established in 1964, under the charter of the National Academy of Sciences, as a parallel organization of outstanding engineers. It is autonomous in its administration and in the selection of its members, sharing with the National Academy of Sciences the responsibility for advising the federal government. The National Academy of Engineering also sponsors engineering programs aimed at meeting national needs, encourages education and research, and recognizes the superior achievements of engineers. Dr. Robert M. White is president of the National Academy of Engineering.

The Institute of Medicine was established in 1970 by the National Academy of Sciences to secure the services of eminent members of appropriate professions in the examination of policy matters pertaining to the health of the public. The Institute acts under the reponsibility given to the National Academy of Sciences by its congressional charter to be an adviser to the federal government and, upon its own initiative, to identify issues of medical care, research, and education. Dr. Samuel O. Thier is president of the Institute of Medicine.

The National Research Council was organized by the National Academy of Sciences in 1916 to associate the broad community of science and technology with the Academy's purposes of furthering knowledge and advising the federal government. Functioning in accordance with general policies determined by the Academy, the Council has become the principal operating agency of both the National Academy of Sciences and the National Academy of Engineering in providing services to the government, the public, and the scientific and engineering communities. The Council is administered jointly by both Academies and the Institute of Medicine. Dr. Frank Press and Dr. Robert M. White are chairman and vice chairman, respectively, of the National Research Council.

Support for this project was provided by grants from the Agricultural Research Service of the U.S. Department of Agriculture and by contributions from the Foundation for Agronomic Research and the Richard Lounsberry Foundation. It also has received support from the National Research Council Fund, a pool of private, discretionary, nonfederal funds that is used to support a program of Academy-initiated studies of national issues in which science and technology figure significantly. The NRC Fund consists of contributions from a consortium of private foundations including the Carnegie Corporation of New York, the Charles E. Culpeper Foundation, the William and Flora Hewlett Foundation, the John D. and Catherine T. MacArthur Foundation, the Andrew W. Mellon Foundation, the Rockefeller Foundation, and the Alfred P. Sloan Foundation; the Academy Industry Program, which seeks annual contributions from companies that are concerned with the health of U.S. science and technology and with public policy issues with technological content; and the National Academy of Sciences and the National Academy of Engineering endowments.

Library of Congress Cataloging-in-Publication Data

National Research Council (U.S.). Committee on a
    National Strategy for Biotechnology in Agriculture.
    Agricultural biotechnology.

    Bibliography: p.
    Includes index.
    1. Agricultural biotechnology—United States.
2. Agricultural biotechnology—Government policy—United
States. 3. Agriculture—Research—United States.
4. Agricultural—Research—United States. I. Title.
S494.5.B563N37 1987   631.3'0973   87-12181
ISBN 0-309-0-3745-X

Printed in the United States of America

Preface

The breakthroughs in science that permitted genes, and thus heredity, to be identified and manipulated as molecules ushered in the biotechnology era, which is now more than a decade old. The new tools of biotechnology are changing the way scientists can address problems in the life sciences; agriculture is one area facing major changes as a result of this new technology. The unanticipated rapid rate at which discoveries and their applications in biotechnology have unfolded has stressed the capacity of society—more specifically, our agricultural research and educational institutions—to absorb and adjust to change. We are challenged by pressing decisions, opportunities, and problems that we face now and will continue to face in the future. Competition from abroad impels us to devise and use new technologies that can improve the efficiency and quality of U.S. agricultural production. These concerns led to this study—an overview of how the agricultural research system is responding to biotechnology and how it might prepare for future opportunities.

The Board on Agriculture initiated this study to explore ways of accelerating the benefits of biotechnology within the U.S. agricultural economy. Support was sought from the National Research Council Fund and the U.S. Department of Agriculture, which also requested a study of public and private sector interactions in biotechnology research. Our committee was asked to examine the activities and issues that biotechnology was generating in research and practical applications, and to recommend strategies by which agriculture might respond to and benefit from these changes. Specifically, the mandate to our committee was to assess

- applications of biotechnology for improving the efficiency of agricultural practice;
- the capacity of existing institutions and programs to train and retrain scientists and carry out research in agricultural biotechnology;
- models and approaches for fostering interdisciplinary research combining the interests and talents of molecular biologists with those of scientists in traditional agricultural disciplines; and
- the role of new interactions for scientific exchange and technology transfer between the private sector and publicly supported research and educational institutions.

Biotechnology is moving in many directions with positive results—crop improvement, vaccine development, and diagnostic methods are some impending applications—but the development of biotechnology's tools can be found in almost every agricultural discipline. Advances are confined more by the limits of our knowledge of the agricultural organisms we want to work with and the resources and trained scientists available than by the power of the tools biotechnology provides.

Chapter 1 provides a summary of our findings that includes recommendations aimed at improving support for the integration of biotechnology's tools into agriculture. Chapter 2 introduces the significant uses of these tools in research and discusses some applications pertinent to agriculture. Additional scientific details on gene transfer methods applicable to agricultural organisms are provided in the Appendix.

The remaining three chapters focus predominantly on policy. Chapter 3 reviews the mandate and organization of institutions that carry out or support agricultural research, how agricultural research is funded, and the present role of biotechnology in agricultural research policy. Chapter 4 covers the training of scientists who will utilize the tools of biotechnology in agricultural research. Last, Chapter 5 addresses technology transfer aimed toward bringing the benefits of agricultural biotechnology to the marketplace. Here the report reviews the rapidly changing scene of university, industry, and government interactions concerning new research agreements as well as patent policies. The committee also addresses new roles for agricultural extension and the need for government to rapidly address the regulatory problem of field testing genetically engineered organisms.

Within the past few years the popular press has captured the public's attention with the role biotechnology will play in agriculture, citing both its positive and negative aspects, whether realistic or wildly speculative. As a committee we profess no special insight into what the future will bring, but we do know that the tools of biotechnology will provide the means to better understand the world we live in and thereby increase our knowledge and ability to make wiser decisions.

Charles E. Hess
Chairman

Acknowledgments

The committee wishes to express its gratitude to the many individuals at public and private institutions who generously contributed information crucial to this study. We particularly wish to acknowledge those who responded to our invitation to come to one of our meetings in Washington, D.C. and share their knowledge and insights into agricultural biotechnology. They include Winston J. Brill, Peter R. de Bruyn, Philip Filner, Gordon G. Hammes, Ralph W. F. Hardy, Virginia H. Holsinger, Theodore L. Hullar, Robert J. Kalter, Edgar L. Kendrick, Gretchen S. Kolsrud, Gwen G. Krivi, Robert Nicholas, Mark L. Pearson, Robert Poling, Leroy Randall, M. Howard Silverstein, Gerald Still, Zachary S. Wochok, and J. Gregory Zeikus. The committee gratefully acknowledges the contributions of its consultants, Chris Elfring, Nancy Heneson, and William Magrath, in gathering and organizing material for this report, and Phyllis B. Moses for the background paper on gene transfer methods that she prepared during her tenure in 1985 as an NRC fellow. We have included this paper as an appendix to our report.

The committee also wishes to thank Aida Neel and Susanne Mason, Administrative Secretaries to the study. Finally, the committee appreciates the role of Board on Agriculture staff members James E. Tavares, Project Officer, and Phyllis B. Moses, Staff Officer, in assimilating and expressing our findings in the final report.

Contents

AGRICULTURAL
BIOTECHNOLOGY

1
Executive Summary and Recommendations

A national strategy for biotechnology in agriculture must focus on solving important scientific and agricultural problems, effectively using the funds and institutional structures available to support research, training researchers in new scientific areas, and efficiently transferring technology. This report assesses the status of biotechnology in agricultural research and suggests approaches toward a more effective national strategy for biotechnology in agriculture. Thus far, government at both the state and federal levels has responded with short-term, ad-hoc management approaches; it has not addressed the long-term needs and policy concerns of integrating biotechnology into agricultural research and technology. Short-term management approaches jeopardize the fragile U.S. competitive advantage in biotechnology. Such approaches and uncertain funding create an environment that does not attract the best minds to agricultural research. This report points to policy changes that are needed in funding patterns and in the operation and organization of agricultural research institutions.

THE INTERNATIONAL DIMENSION

Agriculture has moved from a resource-based to a science-based industry as science and technology have been substituted for land and labor. This transition, which began in the United States, now affects agriculture and food producing systems throughout the world. Technology has driven this change toward more effective and efficient production practices. Yet current political and economic policies governing agriculture neither fully recognize nor take these changes into account. The adoption of new technologies has improved the efficiency of agricultural production practices; the causes of current agricultural surpluses lie elsewhere. Agricultural systems throughout the world continue to adopt new and better technologies that enable them to become more efficient and competitive in developing new markets and capturing old markets for their agricultural products. The future leadership and competitiveness of the U.S. agricultural enterprise is dependent on the health and effectiveness of the agricultural research system in our country and its ability to translate better technologies into practice. Research must make American farming a more profitable, reliable, and durable business able to compete in both domestic and international markets. Innovation is crucial to enhance productive efficiency and environmental acceptability. Biotechnology is key to this innovation.

American agriculture has achieved its preeminence through innovation and substitution of knowledge for resources. This trend must continue. Yet technological innovations cannot revitalize American agriculture unless farm business management, farm policy, the U.S. Department of Agriculture (USDA), land-grant universities, extension services, and private sector businesses that serve agriculture are innovative.

Leadership in technology development and utilization is the role the United States has, can, and should play for the world. American farmers can take the lead in adopting new biotechnologies. These technologies should emphasize maximizing economic yield rather than total production. That is, they should increase the efficiency of production by reducing the costs of production. Such technologies become increasingly important as support prices are removed and world competition stiffens.

A focus on increasing profit by reducing costs requires augmenting our knowledge in the agricultural sciences, especially those fundamental disciplines that underlie biotechnology development. For example, how can one design crops that grow more efficiently and yield more nutritional food? Research will open the door to ever better technologies and products. In both research and development, our USDA laboratories, land-grant universities, and other public and private institutions have a critical role to play.

THE POWER OF BIOTECHNOLOGY

The power of biotechnology is no longer fantasy. Biotechnology—the use of technologies based on living systems to develop commercial processes and products—now includes the techniques of recombinant DNA, gene transfer, embryo manipulation and transfer, plant regeneration, cell culture, monoclonal antibodies, and bioprocess engineering. Using these techniques, we have begun to transform ideas into practical applications. For instance, scientists have learned to genetically alter certain crops to increase their tolerance to certain herbicides. Biotechnology has also been used to develop safer vaccines against viral and bacterial diseases such as pseudorabies, enteric colibacillosis (scours), and foot-and-mouth disease. Yet we have barely scratched the surface of the many potential benefits the tools of biotechnology will bring.

Biotechnology offers new ideas and techniques applicable to agriculture. It offers tools to develop a better understanding of living systems, of our environment, and of ourselves. Yet continued advances will take a serious commitment of talent and funds.

Biotechnology offers tremendous potential for improving crop production, animal agriculture, and bioprocessing. It can provide scientists with new approaches to develop higher yielding and more nutritious crop varieties, improve resistance to diseases and adverse conditions, or reduce the need for fertilizers and other expensive agricultural chemicals. In animal agriculture, its greatest immediate potential lies in therapeutics and vaccines for disease control. Bioprocessing—the use of living systems or their components to create useful products—offers opportunities to manufacture new products and foods, treat and use wastes, and use renewable resources for fuel. Biotechnology could also improve forestry and its products, fiber crops, and chemical feed stocks.

STRATEGIES FOR NATIONAL COMPETITIVENESS

It is important to develop a national strategy for biotechnology in agriculture because biotechnology offers opportunities for increased sustainability, profitability, and international competitiveness in agriculture. Such a strategy should address improving the full spectrum of activities, from the quality and direction of research to the realization of the benefits of this research in agricultural production.

Research Emphasis

The potential benefits of biotechnology will not be realized without a continued commitment to basic research. Six research areas merit emphasis.

1. Gene identification—locating and identifying agriculturally important genes and creating chromosome maps.
2. Gene regulation—understanding the mechanisms of regulation and expression of these genes and refining the methods by which they may be genetically engineered.
3. Structure and function of gene products—understanding the structure and function of gene products in metabolism and the development of agriculturally important traits.
4. Cellular techniques—developing and refining techniques for cell culture, cell fusion, regeneration of plants, and other manipulations of plant and animal cells and embryos.
5. Development in organisms and communities—understanding the complex physiological and genetic interactions and associations that occur within an organism and between organisms.
6. Environmental considerations—understanding the behavior and effect of genetically engineered organisms in the environment.

The Research System

Funding and institutions provide the foundation for progress in biotechnology. A long-term commitment of adequate support is critical because biotechnology requires a substantial initial investment to acquire and build upon basic knowledge. Applying biotechnology to agriculture will put new demands on existing relationships among research institutions, will influence patterns of funding, and will alter established pathways between research discoveries and commercial developments.

The USDA and the land-grant university system have been the keystones of our national agricultural research system, and they will continue to play an important role in developing biotechnologies. Yet the emergence of biotechnology has brought a variety of new actors—in particular, non-land-grant universities and private companies—into agriculturally related research. An alliance is emerging between public sector basic science and private sector technology development, which should be exploited and enhanced in the area of biotechnology.

A variety of federal, state, and private institutions support agriculturally relevant research. The current total annual expenditure for agricultural research by these institutions is roughly $4 billion. Private industry spends about $2.1 billion annually, mostly on proprietary technology development. The federal–state agricultural research system spends about $1.9 billion annually.

The current agricultural research system depends on basic research, applied research, technology development, and technology transfer (which includes extension). Basic and applied research overlap in biotechnology to perhaps a greater extent than in traditional areas of agricultural science. In realigning the system to promote biotechnology, communication is essential among basic researchers, applied researchers, and farmers and private companies, the end users of technology. For the agricultural research system to be most effective, links among the disciplines of science that support agriculture as well as links between basic and applied research and technology development and transfer must be strengthened.

Peer review must be a key component of any step taken to strengthen and improve the agricultural research system. Peer review, which in its broadest form is also called merit review, is one of the most effective mechanisms available to ensure that federal dollars are invested in high-quality research and that judgments made in allocating research funds are equitable and discerning.

Careful attention to the objectivity, quality, and breadth of expertise represented on review panels is necessary to ensure sound decisions.

New Talent

Implementing advances in biotechnology in agriculture will require a work force of highly skilled scientists who can apply molecular biological techniques to agricultural problems. Because biotechnology research spans a continuum from basic science through practical application, its practitioners must be conversant with the general biology of an organism, with the biochemical and genetic details of its life cycle, and with the needs of modern agriculture.

There is an increasing demand for scientists competent with the tools of biotechnology in academic, government, and industrial laboratories. Yet insufficient federal training programs exist to fulfill these needs, and the few programs currently in place are continually in jeopardy because of budget cuts. Increased federal support for graduate education and postdoctoral training in relevant areas is necessary to ensure the supply of scientists. Four types of programs merit increased federal support: pre- and post-doctoral fellowships, training grants, career development awards, and retraining opportunities. As with other effective national programs, these should be administered on a peer-reviewed, competitive basis.

Applications and Commercialization

The goal of technology transfer has always been implicit in U.S. science policy: Federally funded research should benefit the public, and such benefit includes the development and transfer of technologies from public laboratories to private industry. Translating basic research discoveries into commercial applications and social benefits requires a complex set of interactions involving many types of people and institutions. Universities as well as state and federal agencies are expanding their relationships with the private sector as they explore ways to increase scientific communication and the flow of technology.

The rapid rate of breakthroughs in molecular biology and biotechnology and their potential commercial applications have led to more formal and aggressive transfer of biotechnology. The shift has promoted collaborative research relationships between publicly supported scientists in universities and federal laboratories and those in the private sector. Consultancies, affiliate programs, grants, consortia, research parks, and other forms of partnership between the public and private sectors foster communication and technology transfer.

The scientific advances that made biotechnology possible came out of basic research funded mainly by the federal government and carried out primarily at universities. Research in other nations has also made valuable contributions to this area of science. In contrast, industry's support for basic research is quite limited and cannot be expected to compensate for a reduction in federal funding. Thus, continued research efforts at universities remain highly dependent on federal and state governments for support.

Patenting, licensing, and regulatory issues are all areas that affect the rate and cost of technology transfer. In agricultural biotechnology, technology transfer has been hindered by federal government delays in implementing a mechanism to regulate environmental testing of the products of biotechnology. Although patent policy has been modified at the federal level to overcome obstacles that had kept government- and university-sponsored research from being commercially exploited, many government and university institutions retain policies that inhibit the transfer of technology to industry. Of particular importance in technology transfer from federal laboratories will be implementation of the Federal Technology Transfer Act of 1986.

RECOMMENDATIONS

Biotechnology offers both challenge and tremendous opportunity. The Committee on a National Strategy for Biotechnology in Agriculture recommends the following actions as constructive steps in developing and implementing a strategy to utilize biotechnology to improve U.S. competitiveness in agricultural production. Such a strategy addresses not only the science aspects of biotechnology, but also the policy areas of funding and institutions, training, and technology transfer.

Scientific Aspects

INCREASED EMPHASIS ON BASIC RESEARCH

Basic research programs in physiology, biochemistry, genetics, and molecular biology within agricultural disciplines such as agronomy, entomology, and animal science need to be strengthened and in many cases redirected to questions of identifying genes and understanding the regulation of their expression. Just as an enormous information base has provided a substructure for sweeping advances in biomedical science, a similar foundation of knowledge is now needed about the basic biochemistry, physiology, and genetics of such agricultural subjects as host–pathogen interactions, plant and animal developmental responses to environmental stimuli, enzymes and metabolic pathways, and molecular constituents and their patterns of organization in subcellular organelles. Acquiring such knowledge will affect the rate at which agriculturally valuable genes can be identified, isolated, and characterized, and is a prerequisite for applying the tools of biotechnology to agricultural problems.

IMPROVED TECHNIQUES AND APPLICATIONS

The repertoire of molecular biology and cell culture techniques needed to implement advances in genetic engineering is incomplete. Methods for gene transfer in many plants, animals, and microbes; plant cell culture and regeneration; and animal embryo culture and manipulation are inadequate to support the goal of improving agricultural productivity. Increased efforts are needed to apply techniques developed for laboratory organisms to those plants, animals (including insects), and microbes relevant to agriculture.

A national effort should be mounted by both public and private sectors to apply techniques of biotechnology to problems in the agricultural sciences. This effort should include research on:

- Gene identification—locating and identifying agriculturally important genes and creating chromosome maps.
- Gene regulation—understanding the regulation and expression of these genes and refining methods by which they may be genetically engineered.

- Structure and function of gene products—studying the structure and function of gene products in metabolism and the development of agriculturally important traits.
- Cellular techniques—developing and refining techniques for cell culture, cell fusion, regeneration of plants, and other manipulations of plant and animal cells and embryos.
- Development in organisms—using the new technology to study cell and organismic biology in intact organisms.
- Development in communities—understanding the complex associations and interactions that occur among organisms.

INCREASED ATTENTION TO THE ECOLOGICAL ASPECTS OF BIOTECHNOLOGY

Both the public and private sectors should increase their efforts to develop an extensive body of knowledge of the ecological aspects of biotechnology in agriculture. In particular, studies must be done to further our understanding of the behavior and effects of genetically engineered organisms. In addition, the public must be educated about biotechnology. These efforts are essential to support future applications of biotechnology and to adequately inform regulators and the public about both the benefits and possible risks involved.

Funding and Institutions

LINKING AND INTEGRATING RESEARCH

The tools and approaches of biotechnology are equally relevant to science-oriented research and technology-oriented research. Biotechnology can strengthen as well as benefit from improved linkages between basic scientific research and research to adapt technology to agricultural problems. Equally important, different disciplines within biology and agriculture can collaborate to integrate knowledge and skills toward new advances in agriculture.

New approaches to agricultural research are needed to establish productive linkages between basic science and its applications as well as interdisciplinary systems approaches that focus a number of skills on a common mission. Just as biochemistry, genetics, molecular biology, and fields of medicine have successfully joined forces to solve medical problems, integration of these scientific disciplines for agricultural research must be promoted and supported by appropriate recognition and reward through university, industry, and government channels.

First, universities should establish graduate programs that cut across departmental lines; recognize and reward faculty contributions to cooperative research programs; promote collaborative projects and exchanges between researchers in land-grant universities, non-land-grant universities, industry, and government laboratories; and recruit faculty to create interdisciplinary research programs that can attract competitive funding. Faculty should be selected by departments or groups representing two or more disciplines (e.g., genetics and entomology or biochemistry and botany).

Second, federal and state governments should support the establishment of collaborative research centers, promote interdisciplinary conferences and seminars, support sabbaticals for government scientists and other exchange and retraining programs with universities and industrial laboratories, and provide funding for interdisciplinary-program project grants.

PEER AND MERIT REVIEW

A peer and merit review process must be used to assess and guide the development of the agricultural biotechnology research system, including all steps from basic science to extension.

The participants and procedures in the review process should be organized to match the nature of the tasks and programs reviewed and must include individuals outside the organization as well as experts from relevant disciplines and from basic and applied research programs.

Efforts must be made to broaden the expertise represented on review panels in order to best examine the quality and relevance of work with minimal bias. The benefits of peer and merit review— properly done and heeded—are continuous monitoring of research advances; more efficient, relevant, and higher quality research; and increased communication and respect among scientists.

THE FEDERAL GOVERNMENT'S ROLE

It is logical that primary funding for agricultural biotechnology should be achieved through the USDA. Unfortunately, funding for both intramural and extramural basic research within USDA is well below that of other federal agencies. USDA has recognized the need to support basic research and is attempting to do so, albeit not as rapidly as might be optimal. Funding increases are needed. Allocation of new and even redirected funding should be based principally on competitive peer and merit review.

Any increase in funding at USDA should not come at the expense of appropriations to other federal agencies that support biological research relevant to agriculture. This is because it is not always clear where innovation applicable to agricultural biotechnology might arise. However, some existing research program funds should be redirected within USDA to heighten the priority given to biotechnology. USDA should also emphasize related fundamental research on animals and plants, the lack of which is impeding the application of biotechnology to livestock and crop improvement.

Funding for competitive grants through USDA must be of a size and duration sufficient to ensure high-quality, efficient research programs. The recommended average grant should be increased to $150,000 per year for an average of 3 years or more. This level of funding is consistent with the current average support per principal investigator used by industry and the USDA's Agricultural Research Service (ARS) intramural research programs. The duration of these competitive grants is also in accord with the recent recommendation:

Of equal importance with the level of funding is the stabilization of federal support to permit more effective use of financial and human resources. . . . Federal agencies [should] work toward an average grant or contract duration of at least three, and preferably five, years. (White House Science Council, 1986)

The committee recommends that competitive grants by all agencies in the federal government for biotechnology research related to agriculture total upwards of $500 million annually, a level that could support 3,000 active scientists. This level of support should be achieved by 1990, primarily through competitive grants administered by USDA and the National Science Foundation.

THE STATE GOVERNMENTS' ROLE

States should continue to strengthen their already major role in agricultural research and training through their support of universities and research stations that conduct regional research. They should continue to focus on identifying regional interests and on supporting the training of personnel needed in agriculture. The states should also evaluate programs in agricultural biotechnology and the role such programs can and will play in each state's economy.

THE PRIVATE SECTOR'S ROLE

The private sector's traditional emphasis on product development is not likely to change, even though there has been a dramatic increase since 1980 in private sector investment in high-risk basic research in agricultural biotechnology. Because public sector investment provides skilled manpower and the knowledge base for innovation, industry should act as an advocate for publicly supported training and research programs in agricultural biotechnology. Industry can also support biotechnology research through direct grants and contracts to universities, cooperative agreements with federal laboratories, and education to inform the general public about the impacts of agricultural biotechnology.

Foundations should be encouraged to support innovative science programs in order to maximize their potential for having substantial influence in important areas. The McKnight Foundation's interdisciplinary program for plant research and the Rockefeller Foundation's efforts to accelerate biotechnology developments in rice are noteworthy examples. Other foundations should address equally important experiments in technology transfer and extension for agricultural biotechnology.

Training

Scientists, administrators, faculty, and policymakers in all sectors should be aware of the importance of state-of-the-art education and training to the future development of agricultural biotechnology. Specifically, the committee makes the following recommendations.

INCREASED FEDERAL SUPPORT FOR TRAINING

Major increases in federal support for training programs are urgently needed to provide a high-quality research capability that ensures the future of U.S. agriculture and meets the growing need for scientists trained in agricultural biotechnology. Four types of programs must be supported: pre- and postdoctoral fellowships, training grants, career development awards, and retraining opportunities. These approaches, used successfully in the biomedical sciences, have put the United States in the forefront of human medical advances. These programs should be administered on a peer-reviewed, competitive basis. USDA should support at least 400 postdoctoral positions at universities and within the ARS, which represents a quadrupling of the present number, and maintain strong support for graduate-level training.

INCREASED RETRAINING PROGRAMS

For the short term, highest priority should go to increasing the retraining opportunities available to university faculty and federal scientists to update their background knowledge and provide them with laboratory experience using the tools of biotechnology. This retraining will expand the abilities of researchers experienced in agricultural disciplines. USDA should take the lead in administering a program to supply at least 150 retraining opportunities a year for 5 years, starting in FY89.

Technology Transfer

ROLES FOR UNIVERSITIES AND GOVERNMENT AGENCIES

Universities and state and federal agencies are expanding both the nature and number of their relationships with the private sector as they explore ways to increase scientific communication and the flow of technology. The federal government, granting agencies, and public and private universities should encourage interdisciplinary research, partnerships, and new funding arrangements among universities, government, and industry. The Federal Technology Transfer Act of 1986 provides new incentives to federal scientists in this regard. Consultancies, affiliate programs, grants, consortia, research parks, and other forms of partnership between the public and private sectors that foster communication and technology transfer should be promoted. The USDA, State Agricultural Experiment Stations, and the Cooperative Extension Service (CES) should emulate other agencies such as the National Institutes of Health and the National Bureau of Standards in forming innovative affiliations to increase technology transfer.

COOPERATIVE EXTENSION SERVICE

The CES should focus some of its efforts on the transfer of biotechnology research that will prove adaptable and profitable to the agricultural community. It should train many of its specialists in biotechnology and increase its interactions with the private sector to keep abreast of new biotechnologies valuable to the agricultural community. Furthermore, CES should work to anticipate and alleviate social and economic impacts that may result from the application of biotechnologies. CES should also play a key role in educating the public about biotechnology.

PATENTING AND LICENSING

Patenting and licensing play necessary roles in advancing technology transfer and assuring the commercialization of research results, especially in capital-intensive fields such as biotechnology. Patenting and licensing by universities and government agencies should be encouraged as key instruments used to transfer technology. Universities and government agencies should provide incentives to their scientists to encourage patenting. Public policy should encourage state land-grant universities to confer exclusive licenses on patents to private companies with the resources, marketing, and product interests required to translate these discoveries into commercial products.

REGULATION OF FIELD TESTING

The government's uncertainty over appropriate regulatory steps has fueled public controversy over the assessment of possible environmental risks from genetically engineered agricultural products. USDA, the Food and Drug Administration, and the Environmental Protection Agency must formulate, publish, and implement a research and regulatory program that is based on sound scientific principles. Initially, 5–10 selected, already-existing publicly owned field stations should be available as an option for environmental release testing, professionally managed by an oversight committee of public sector scientists with expertise in agronomy, ecology, plant pathology, entomology, microbiology, molecular biosciences, and public health. This interim program should be designed to gain scientific information and practical experience with field testing and to protect the public safety. The current lack of adequate regulatory procedures is halting progress in applying biotechnologies to agriculture.

2
Scientific Aspects

THE POWER OF BIOTECHNOLOGY

The tools of biotechnology offer both a challenge and tremendous opportunity. They do not change the purpose of agriculture—to produce needed food, fiber, timber, and chemical feed stocks efficiently. Instead, they offer new techniques for manipulating the genes of plants, animals, and microorganisms. Biotechnology tools complement, rather than replace, the traditional methods used to enhance agricultural productivity and build on a base of understanding derived from traditional studies in biology, genetics, physiology, and biochemistry.

Biotechnology has opened an exciting frontier in agriculture. The new techniques provided by biotechnology are relatively fast, highly specific, and resource efficient. It is a great advantage that a common set of techniques—gene identification and cloning, for example—are broadly applicable. Not only can we improve on past, traditional methods with the more precise modern methods, but we can explore new areas as well. We can seek answers to questions that only a few years ago we never thought to ask.

The power of biotechnology is no longer fantasy. In the last few years, we have begun to transform ideas into practical applications. For instance, scientists have learned to genetically alter certain crops to increase their tolerance to certain herbicides. Biotechnology has been used to design and develop safer and more effective vaccines against viral and bacterial diseases such as pseudorabies, enteric colibacillosis (scours), and foot-and-mouth disease.

Yet we have barely scratched the surface of the potential benefits. Much remains to be learned, and continued advances will take a serious commitment of talent and funds (see Chapter 3).

This chapter briefly reviews the major uses of biotechnology in agriculture. It looks specifically at the progress and potentials of genetic engineering and other new biotechnologies in plant and animal agriculture and bioprocessing. These sections review traditional approaches, discuss examples of progress using biotechnology, and highlight opportunities on the horizon.

USING GENE TRANSFER TO ENHANCE AGRICULTURE

Throughout the history of agriculture, humans have taken advantage of the natural process of genetic exchange through breeding that creates variation in biological traits. This fact underlies all attempts to improve agricultural species, whether through traditional breeding or through techniques of molecular biology. In both cases, people manipulate a natural process to produce varieties of organisms that display desired characteristics or traits, such as disease-resistant crops or food animals with a higher proportion of muscle to fat.

The major differences between traditional breeding and molecular biological methods of gene transfer lie neither in goals nor processes, but rather in speed, precision, reliability, and scope. When traditional breeders cross two sexually reproducing plants or animals, tens of thousands of genes are mixed. Each parent, through the fusion of sperm and egg, contributes half of its genome (an organism's entire repertoire of genes) to the offspring, but the composition of that half varies in each parental sex cell and hence in each cross. Many crosses are necessary before the "right" chance recombination of genes results in offspring with the desired combination of traits.

Molecular biological methods alleviate some of these problems by allowing the process to be manipulated one gene at a time. Instead of depending on the recombination of large numbers of genes, scientists can insert individual genes for specific traits directly into an established genome. They can also control the way these genes express themselves in the new variety of plant or animal. In short, by focusing specifically on a desired trait, molecular gene transfer can shorten the time required to develop new varieties and give greater precision. It also can be used to exchange genes between organisms that cannot be crossed sexually.

Gene transfer techniques are key to many applications of biotechnology. The essence of genetic engineering is the ability to identify a particular gene—one that encodes a desired trait in an organism—isolate the gene, study its function and regulation, modify the gene, and reintroduce it into its natural host or another organism. These techniques are tools, not ends in themselves. They can be used to understand the nature and function of genes, unlock secrets of disease resistance, regulate growth and development, or manipulate communication among cells and among organisms.

Isolation of Important Genes

The first step in an effort to genetically engineer an organism is to locate the relevant gene(s) among the tens of thousands that make up the genome. Perhaps the researcher is searching for genes to improve tolerance to some environmental stress or to increase disease resistance. This can be a difficult task—similar to trying to find a citation in a book without an index.

This task is made easier with restriction enzymes that can cut complex, double-stranded macromolecules of DNA into manageable pieces. A restriction enzyme recognizes a unique sequence in the DNA, where it snips the strands. By using a series of different restriction enzymes, an organism's genomic DNA can be reduced to lengths equivalent to one or several genes. These smaller segments can be sorted and then cloned to produce a quantity of genetic material for further analysis. The collection of DNA segments from one genotype—a gene library—can be searched to locate a desired gene. Patterns can also be analyzed to link a particular sequence—a marker—to a particular trait or disease, even though the specific gene responsible is still unknown.

Restriction enzymes are also used in cloning genes. To clone a gene, a small circle of DNA that exists separate from an organism's main chromosomal complement—a plasmid—is cut open using the same restriction enzyme that was used to isolate a desired gene. When the cut plasmid and the isolated gene are mixed together with an enzyme that rejoins the cut ends of DNA molecules, the isolated gene fragment is incorporated into the plasmid ring. As the repaired plasmid replicates, the cloned gene is also replicated. In this way, numerous reproductions of the cloned gene are produced within the host cell, usually a bacterium. After replication, the same restriction enzyme is used to snip out the cloned gene, allowing numerous copies of that gene to be isolated.

The ability to isolate and clone individual genes has played a critical role in the development of biotechnology. Cloned genes are necessary research tools for studies of the structure, function, and expression of genes. Further, specific gene traits could not be transferred into new organisms unless numerous gene copies were available. Cloned genes also are used as diagnostic test probes in medicine and agriculture to detect specific diseases.

Gene Transfer Technology

To transfer genes from one organism to another, molecular biologists use vectors. Vectors are the "carriers" used to pass genes to a new host, and they can mediate the entry, maintenance, and expression of foreign genes in cells. Vectors used to transfer genes include viruses, plasmids, and mobile segments of DNA called transposable elements. Genes can also be introduced by laboratory means, such as chemical treatments, electrical pulses, and physical treatments including injection with microneedles. The basic principles behind these technologies are the same for animals, plants, and microbes, although specific modifications may be necessary. (The basic gene transfer methods are described in detail in the Appendix, "Gene Transfer Methods Applicable to Agricultural Organisms.")

Vectors based on viruses, plasmids, and transposable elements have been adapted from naturally occurring systems and engineered to transfer desired genes into animals, plants, and microbes. For plants, the classic example is the Ti plasmid from the soil bacterium *Agrobacterium tumefaciens*, which in nature transfers a segment of DNA into plant cells, causing the recipient cells to grow into a tumor. Scientists have adapted this plasmid by eliminating its tumor-causing properties to create a versatile vector that can transfer foreign genes into many types of plants.

Similarly, the transposable P-element of the fruit fly *Drosophila melanogaster* is an effective vector for gene transfer into *Drosophila*. This or similar transposable elements should prove to be adaptable to insects of agricultural importance. Animal viruses such as simian virus 40 (SV40), adeno, papilloma, herpes, vaccinia, and the retroviruses, all originally studied because of their role in disease, are now being engineered as vectors for gene transfer into animal cells and embryos. Plant viruses such as cauliflower mosaic virus, brome mosaic virus, and geminiviruses are similarly being exploited for their abilities to transfer genes.

Cell Culture and Regeneration Techniques

The ability to regenerate plants from single cells is important for progress with gene transfer into plants. Animals cannot be regenerated asexually, so the only way to introduce a foreign gene into all cells of an animal is to insert it into the sperm, egg, or zygote. Cell culture techniques are important for the regeneration of plants. They are also critical for fundamental studies on both plant and animal cells, and for the manipulation of microorganisms.

The vegetative propagation of stem cuttings or other growing plant parts to produce genetic clones is common for some agricultural crops. Potatoes, sugarcane, bananas, and some horticultural species, for example, are cultivated by vegetative propagation. Techniques exist to propagate and regenerate whole plants from tissues, isolated plant cells, or even protoplasts (plant cells from which the cell wall has been enzymatically removed) in culture. This set of techniques is complete for some agricultural species, such as alfalfa, carrots, oilseed rape, soybeans, tobacco, tomatoes, and turnips. Progress on other crops, including major food species such as many cereals and legumes, has been slower.

Cell culture techniques have taken on added importance as biotechnology has progressed. Genetic engineering requires an ability to manipulate individual cells as recipients of isolated genes. Cell culture techniques allow scientists to maintain and grow cells outside the organism and thus expand their ability to perform gene transfer and study the results. In addition, cell culture allows scientists to regenerate numerous copies (clones) of the manipulated varieties, which is easier, more efficient, and more convenient, especially for producing significant quantities of stock plants. A third use of cell culture is to regenerate "somaclonal variants," plants with altered genetic traits that can prove useful as new or improved crops. Thus, cell culture techniques are important to increasing the productivity and versatility of agriculture.

However, there are some important limitations. Chromosomal abnormalities appear as cultures age. These changes are related to the phenomenon of somaclonal variation, which may prove useful to agriculture, but in many instances the changes are undesirable. Therefore, scientists must learn how to prevent chromosomal changes in cell cultures. Second, long-term cultures lose regenerative potential. As biotechnology expands, it will be critical to understand why different species have differing abilities to regenerate from cell cultures into plants and how factors such as the genetic or physiological origin of the cells and the culture conditions affect growth. Most plant cells appear to be totipotent, that is, they are in a reversible differentiated state that will permit them to regenerate into a whole plant under appropriate conditions. Understanding what these appropriate conditions are remains a fundamental question in the study of plant development and its genetic control.

Monoclonal Antibody Technologies

The development of monoclonal antibody technology is based on advances in our ability to culture cells. Antibodies are the protein components of the immune system found in the blood of mammals. They have a unique ability to identify particular molecules and select them out. When a foreign substance (an antigen) enters the body, specialized cells called B lymphocytes produce a protein (an antibody) to combat it. To envision how antibodies work, think of a lock and key: The antibody key "fits" only the specific antigen lock. This marks the antigen for destruction. Each of the specialized B lymphocyte cells produces only a single type of antibody and thus recognizes only one antigen.

Apart from their natural role in protecting organisms via the immune response, antibodies are important scientific tools. They are used to detect the presence and level of drugs, bacterial and viral products, hormones, and even other antibodies in the blood. The conventional method of producing antibodies is to inject an antigen into a laboratory animal to evoke an immune response. Antiserum (blood serum containing antibodies) is then collected from the animal. However, antiserum collected in this way contains many types of antibodies, and the amount that can be collected is limited.

Modern biotechnology has opened a door to a more efficient, more specific, and more productive way of producing antibodies. By fusing two types of cells, antibody-producing B lymphocytes and quasi-immortal cancer cells from mice, scientists found that the resulting hybrid cells, called hybridomas, secreted large amounts of homogeneous antibodies. Each hybridoma has the ability to grow indefinitely in cell culture and thus can produce an almost unlimited supply of a specific "monoclonal" antibody. By immunizing mice with specific antigens, researchers can create and select hybridomas that produce a culture of specific, desired monoclonal antibodies.

Thus, biotechnology has produced a way of creating pure lines of antibodies that can be used to identify complex proteins and macromolecules. Monoclonal antibodies are powerful tools in molecular analyses, and their uses in detecting low levels of disease agents such as bacteria and viruses are rapidly expanding.

Beyond many diagnostic uses, hybridoma technology shows promise for immunopurification of substances, imaging, and therapy. Immunopurification is a powerful technique to separate large, complex molecules from a mixture of either unrelated or closely related molecules. For imaging, easily visualized tags can be attached to monoclonal antibodies to provide images of organs and to locate tumors to which the antibody will specifically bind. Finally, new therapeutic methods have been developed that use monoclonal antibodies to inactivate certain kinds of immunological cells and tumor cells or to prevent infection by certain microorganisms.

Although many applications of this technology are still in the experimental stages, the commercial agricultural use of monoclonal antibodies has begun. For example, monoclonal antibodies are now on the market as therapeutics against calf and pig enteric colibacillosis, which causes neonatal diarrhea (scours). This approach is often more effective than conventional vaccines, and it supplements genetically engineered vaccines. Monoclonal antibody-based diagnostic kits can detect whether scouring animals are infected with a particular strain of an *Escherishia coli* bacterium that causes scours, and thus help veterinarians determine the appropriate therapeutic monoclonal antibody to use on an infected herd.

Summary

In its simplest form, genetic engineering involves inserting, changing, or deleting genetic information within a host organism to give it new characteristics. This technology will likely bring great benefits to agriculture, just as breeding has over several thousand years of human history. The development and use of new techniques is allowing researchers to manipulate the genetic character of organisms while overcoming the complications and limitations of sexual gene exchange. Genetic engineering is reducing the amount of time needed to analyze genetic information and transfer genes. Both genetic engineering and monoclonal antibody technology, another major development in biotechnology, greatly increase the specificity and accuracy of analytical research methods. Further, these new technologies are permitting highly specific molecular analyses to be done and are opening new areas of inquiry. The tools of biotechnology, combined with traditional techniques in biology and chemistry, increase enormously both the power and the pace of discoveries in biological investigation.

NEW APPROACHES TO CROP PRODUCTION

In the past 50 years, agricultural production in the United States has more than doubled while the amount of land under cultivation has actually declined slightly. This impressive agricultural success is the result of many factors: an abundance of fertile land and water, a favorable climate, a history of innovative farmers, and a series of advances in the science and technology of agriculture that have made possible more intensive use of yield-enhancing inputs such as fertilizer and pesticides. Yet the productivity successes brought about by farm mechanization, improved plant varieties, and the development of agricultural chemicals may be harder to repeat in the future unless new approaches are pursued.

Biotechnology offers vast potential for improving the efficiency of crop production, thereby lowering the cost and increasing the quality of food. The tools of biotechnology can provide scientists with new approaches to develop higher yielding and more nutritious crop varieties, to improve resistance to disease and adverse conditions, or to reduce the need for fertilizers and other expensive agricultural chemicals. The following paragraphs highlight some examples of how genetic engineering can be used to enhance crop production.

The Genetic Engineering of Plants

Perhaps the most direct way to use biotechnology to improve crop agriculture is to genetically engineer plants—that is, alter their basic genetic structure—so they have new characteristics that improve the efficiency of crop production. The traditional goal of crop production remains unchanged: to produce more and better crops at lower cost. However, the tools of biotechnology can speed up the process by helping researchers screen generations of plants for a specific trait or work more quickly and precisely to transfer a trait. These tools give breeders and genetic engineers access to a wider universe of traits from which to select.

Although powerful, the process is not simple. Typically, researchers must be able to isolate the gene of interest, insert it into a plant cell, induce the transformed cell to grow into an entire plant, and then make sure the gene is appropriately expressed. If scientists were introducing a gene coding for a plant storage protein containing a better balance of essential amino acids for human or animal nutrition, for example, it would need to be expressed in the seeds of corn or soybeans, in the tubers of potatoes, and in the leaves and stems of alfalfa. In other words, the expression of such a gene would need to be directed to different organs in different crops.

Putting the New Technologies to Work

There are already successes that demonstrate how plants can be genetically engineered to benefit agriculture. Herbicide resistance traits are being transferred to increase options for controlling weeds. Soon, the composition of storage proteins, oils, and starches in plants may be altered to increase their value.

One plant gene that has been isolated, cloned, and transferred is for the sulfur-rich protein found in the Brazil nut, *Berthalletia excelsa*. This protein contains large amounts of two nutritionally important sulfur-containing amino acids: methionine and cysteine. These are the very nutrients in which legumes, such as soybeans, are deficient. If the sulfur-rich protein gene were transferred into soybeans, it might enhance this legume's role as a protein source throughout the world.

By purifying the Brazil nut protein and determining the order and kind of amino acids in the protein, scientists were able to synthesize an artificial segment of DNA coding for a section of this protein. This DNA "probe" was used to find and pull out the natural gene from the Brazil nut. Researchers then transferred the gene into tomato and tobacco plants, which were chosen because they are easier to manipulate than soybeans. Researchers have also transferred the gene into yeast cells. Early results show that the genetically engineered yeast do produce the sulfur-rich protein.

Similar work is being done to improve oil crops. Oil crops produced in the United States in 1984 were worth $11.8 billion. Depending on their chemical composition, oils and waxes from plants have uses in feed, food, and industrial products such as paints and plastics. Chemical properties, and thereby the uses of plant oils, vary depending on the length of the fatty acid chains that compose the oil and their degree of saturation. Many of the enzymes controlling the biochemical pathways that regulate molecular chain length and degree of saturation have been well studied, and this reservoir of knowledge now makes it possible to genetically engineer the type of oil a crop produces. Although traditional breeding methods have succeeded in modifying the oil composition of some crops, genetic engineering opens a broader range of possibilities.

Scientists have taken another important step in using genetic engineering to improve crop production: They have for the first time engineered plants to be resistant to powerful herbicides. One example is glyphosate (trade name: "Roundup"), a common, effective, and environmentally safe herbicide. However, glyphosate indiscriminantly kills crops as well as weeds. Thus, it must usually be used before crop plants germinate. Yet by engineering crops to be resistant to glyphosate, scientists hope to expand the range of the herbicide's applications.

Scientists have isolated a glyphosate-resistance gene and successfully transferred it into cotton, poplar trees, soybeans, tobacco, and tomatoes. The gene was derived from the bacterium *Salmonella typhimurium*. Similarly to other accomplishments in biotechnology, this success depended on extensive prior basic research on biochemical pathways in bacteria and plants, and sophisticated gene cloning and transfer techniques. Field testing and commercialization of glyphosate-resistant crops should follow soon. Analysis of tomato growers' costs in California predicts that farmers could save up to $100 per acre in weed control costs if they used glyphosate in place of current herbicides, with concomitant reductions in labor, equipment, and environmental damage. This advance would also give farmers improved flexibility, yield, quality, and spectrum of weed control.

LOOKING TO THE FUTURE

With such promising examples already being realized, it is interesting to speculate about other possibilities. For instance, could scientists take naturally occurring chemicals that hinder plant growth—such as the compound crabgrass releases that prevents other grasses from invading its territory—and engineer crop plants with their own ability to control weeds? Scientists have long known that some plants produce chemicals that affect the growth of other plants; by studying these allelopaths, scientists may be able to engineer or breed plants that would give farmers new biological tools to fight weeds, in addition to mechanical cultivation and other cultural tools, and chemical herbicides. The potential value of research on biological methods of weed control is great, but the work is very complicated and significant advances are not expected quickly. One of the complicating factors that must be understood is how certain plants produce allelopathic molecules and at the same time protect themselves against these chemicals.

Observations of nature combined with abilities to engineer plants might also provide opportunities to manipulate plant growth and development. Through research, scientists have determined that flowering, dormancy, fruit-ripening, and a host of other growth and developmental processes come under the influence of a relatively few plant hormones or growth regulating substances.

Agricultural chemists have already discovered a number of inhibitors and mimics of these regulating compounds, and these have readily found commercial applications. For example, they are used to induce and synchronize flowering and fruit production in pineapple fields, to control ripening and premature dropping of fruit from trees and vines, and to block elongation growth to create more compact and attractive potted plants, such as chrysanthemums and poinsettias.

Because the natural growth regulators are active in very small amounts, it has been difficult to study their synthesis and mode of action. However, the availability of new techniques and genetic probes to locate the genes responsible for their synthesis is giving researchers new tools to study these chemicals. As our understanding grows, we will likely discover additional ways to regulate and control plant growth and development. For example, perhaps scientists can improve on ways to control fruit ripening, so ripening can be delayed until the fruit is en route to market. Scientists may also develop ways to increase flowering, fruiting, seed set, or other growth habits of plants to improve efficiency of production.

The Genetic Engineering of Microorganisms
Associated with Plants

Microorganisms in the environment affect the growth of plants in a variety of ways, many of which are still poorly understood. Their effects can be either beneficial or harmful. For instance, some microorganisms protect plants from bacterial or fungal infections. Others protect plants from environmental stresses such as acidity, salinity, or high concentrations of toxic metals. Still others attack weeds that compete with crops. The best known association between microorganisms and plants is the symbiotic relationship between nitrogen-fixing bacteria of the genus *Rhizobium* and members of the legume family, such as soybeans.

However, some microorganisms, particularly certain bacteria and fungi, are pathogens that attack crops and cause disease, sometimes in epidemic proportions. The Irish Potato Famine of the mid-1800s, the Dutch Elm disease of the twentieth century, and the southern corn leaf blight of 1970 are dramatic examples of losses caused by pathogens.

As our understanding of the relationships between microorganisms and crops improves, the genes controlling these relationships—whether in the microorganism or in the plant—can be engineered to enhance the abilities of beneficial microorganisms or inhibit the effects of harmful microorganisms. Yet to successfully engineer microorganisms, scientists must understand the molecular mechanisms by which they interact with their plant hosts. Much remains to be learned about both the plant and microbial genes involved, their regulation, and the intricate relationships between microorganisms and their hosts.

PUTTING THE NEW TECHNOLOGIES TO WORK

Initial discoveries in genetic engineering technologies were made with microorganisms because they are simpler life-forms than higher plants and animals, and thus are easier to manipulate in the laboratory. Methods developed in medical research with bacteria and viruses are now being adapted to agriculturally significant microorganisms. One example that has progressed to the point of field testing involves genetically altered bacteria designed to prevent frost damage. *Pseudomonas syringae* is a bacterial species with many members that are normally harmless and commonly inhabit the outer surface of plant cells. However, some of these bacteria contain a protein that initiates the formation of ice crystals at temperatures below freezing. The growing ice crystals can rupture and damage plant cells. If the bacteria are not present, plants can withstand colder temperatures without damage. Researchers have now created an "ice-minus" strain of *P. syringae* by removing the gene that makes the protein.

In laboratory tests the ice-minus strain has been sprayed on plants to displace the wild strain and thereby provide the crop with some measure of frost protection. Although the genetically engineered, ice-minus *Pseudomonas* is already several years old, field tests necessary to test its commercial application have been blocked by public apprehension that has led to court actions and confusion over the types of precautions needed to regulate such environmental testing.

Another practical application involves the use of DNA probes to detect plant viruses and viroids. Detection permits rapid screening to eliminate infected stock and thus halt the spread of diseases.

Nearly 60 years ago scientists found that a mild strain of tobacco mosaic virus (TMV) could protect tobacco plants against the adverse effects of a subsequently inoculated, severe strain of the virus. This phenomenon, termed cross-protection, has been applied on a limited scale to protect greenhouse tomatoes and a few orchard crops. There are potential problems with the conventional cross-protection approach, however, because the mild, protecting virus might spread to other crops or mutate to a more virulent form. Recently, scientists installed fragments of the TMV genome in tobacco and tomato plants. Because these "transgenic" plants have only a portion of the genetic information that is needed for TMV replication, the problems of conventional cross-protection are avoided. Some transgenic plants appeared to be completely resistant to the TMV virus. Tests show that virus resistance introduced by recombinant DNA technology can be transmitted through seed as a simple Mendelian trait and can thus be transmitted by conventional breeding techniques.

LOOKING TO THE FUTURE

Little is known about the specific genetic and biochemical associations among microorganisms, plants, and the environment, thus many examples of potential changes beneficial to agriculture are still speculative. One area of tremendous promise—genetic engineering to improve nitrogen fixation—is proving particularly challenging.

All living things need nitrogen, yet plants cannot directly absorb and use nitrogen gas, which makes up more than 75 percent of the atmosphere. To be available to plants, nitrogen gas must first be "fixed," or converted into nitrogen-containing compounds either by industrial processes or by certain bacteria and blue-green algae that live in the soil. The most well-known bacteria able to fix nitrogen belong to the genus *Rhizobium*, which associates with members of the legume family such as soybeans, beans, peas, peanuts, alfalfa, and clover. Genetic engineers would like to find ways to improve nitrogen fixation in these plants and extend the ability to others. This development could play a critical role in lowering production costs by reducing the need for energy (petrochemical) inputs used in producing nitrogen fertilizers.

Researchers are pursuing a number of different strategies to improve nitrogen fixation. Perhaps the simplest approach is to improve the symbiotic relationship now found in nature—to genetically engineer *Rhizobium* to fix nitrogen more efficiently for their natural host legumes. A second approach would be to create *Rhizobium* that could infect and fix nitrogen for other plants, in particular the cereal crops. Alternatively, it might be possible to transfer the ability to fix nitrogen to other microorganisms that already live in association with a given crop. Another approach involves trying to engineer plants to fix nitrogen themselves.

Some progress has been made in these approaches, due to extensive basic research on the genetics and biochemistry of nitrogen fixation. Researchers have identified bacterial genes, called *nod* genes, involved in nodulation. When bacteria invade leguminous plants, the *nod* genes are activated, nodules form where the bacteria reside, and nitrogen fixation begins. Researchers are now trying to decipher the chemical signals that activate the bacterium and cause the plant to grow the nodules.

The bacterial genes that actually carry out nitrogen fixation, the *nif* genes, are well studied. Scientists are gaining an understanding of the regulation of these many genes' expression, but their relationship is exceedingly complex. One of many remaining problems is that in the field, laboratory-modified rhizobial inoculants lose out to competing indigenous strains.

Genetic Engineering for Crop Protection

Another strategy to improve crop production through genetic engineering involves protecting crops from pests. Insects, viruses, bacteria, fungi, nematodes, and weeds can all impair agricultural productivity. Yet in a natural ecosystem, organisms typically serve many functions. Insects, for example, can be pests—destroying crops and stored products and transmitting disease. They can also be benefactors—pollinating plants, eating other pests, and recycling organic wastes.

Most chemical insecticides, herbicides, and other pesticides that have been the primary methods of controlling pests are not selective enough to affect only harmful organisms. As biotechnology becomes more refined, methods for handling bothersome pests and beneficial organisms will be created.

PUTTING THE NEW TECHNOLOGIES TO WORK

One area in which genetic engineering technology will prove particularly useful is in developing biological pest control methods. Insects are attracted to certain plants and repelled by others. Some plants produce chemicals that mimic insect hormones and disrupt the reproduction of insects feeding on the plant. Thus, the potential exists to identify the genes controlling the properties and transfer these traits to other plants.

Insect hormones are already used in small quantities in pest management. Pheromones, for example, are used as attractants in traps that monitor levels of insect populations. Conversely, alaromones can be used to repel insects from stored products. Hormones are often structurally complex and their production could require the concerted expression of a number of genes. Thus, extensive basic research on the biosynthetic pathways of these chemicals is necessary before they can be manufactured in microbial, cell culture, or plant systems. Ultimately, as genetic engineers increase their skills, they may be able to alter crops so they produce their own insect repellants.

Some advanced uses of hormones for biological pest control are already available. Juvenile hormone analogues are synthetic chemical compounds similar to a natural hormone that controls maturation in insects. When the juvenile hormone analogue is sprayed on an insect, it remains in an immature state and dies instead of maturing and reproducing. One company that has developed such a substance has registered it with the Environmental Protection Agency (EPA) and is marketing a version for flies, mosquitoes, fleas, and cockroaches. This is a prime example of how knowledge of insect physiology and chemistry can lead to practical applications.

Another experiment of potential importance for insect control involves a genetically altered bacterium. The organism—a strain of corn-root colonizing bacteria called *Pseudomonas fluorescens*— has been genetically changed so it produces an endotoxin that is a potent insecticide for certain pests, including black cutworm. The gene to produce the toxin was transferred from another bacterium, *Bacillus thuringiensis*, which itself has been marketed as a biological insecticide for more than 20 years. The recombinant bacterium can be freeze-dried and coated directly on seeds before planting, or it can be sprayed onto the fields. Tests indicate that the nonrecombinant parental *P. fluorescens* strain remains viable for only 8–14 weeks in the field; then it dissipates and appears to have no long-term effects. Although the current recombinant strain affects a small range of insects, the company developing it intends it to be a prototype for products that could be marketed within the next few years. Successful work at another company has focused on transfer of the toxin gene into plants themselves, which makes them self-protecting against certain insects, notably the tobacco hornworm. In a similar approach, a search is under way for genes controlling resistance or toxins against nematodes.

LOOKING TO THE FUTURE

Naturally occurring insect pathogens, including bacteria, viruses, and fungi, have served for many years as agents of biological pest control, but problems with production, application, and efficacy have prevented their widespread use. However, advances in genetic engineering are opening routes to manipulate these organisms into more useful tools for biological insect control on a large scale.

More than 100 kinds of bacteria have been identified as pathogenic to insects, yet only a few have been examined for their potential to control pests. Pathogenic viruses also hold great potential. Baculoviruses, which are considered inherently safer to work with than other insect viruses because they do not infect vertebrates or plants, seem especially promising. Genetic engineers hope to alter these viruses to produce toxins for specific insects. The virus would infect the insect and then produce the toxin within the insect's cells. Ideally, scientists could design viruses that only harm certain pest species. Baculoviruses are relatively stable in storage, during application, and in the field, and can be produced on a commercial scale. They have been modified with various foreign genes and have expressed those genes in insect cell cultures and silkworm larvae (see the Appendix for details). However, much remains to be learned if scientists are to find appropriate toxin genes.

A more speculative approach to insect control is the use of modified plant viruses that are normally spread by insects. In this strategy an insect-specific toxin gene or behavior-modifying gene would be inserted into the genome of the plant virus, so it is expressed in the cells of the carrier insect. This approach might be a method of controlling sucking insects.

Various fungi, too, are known to cause widespread diseases in insect populations. Most fungal species can penetrate an insect's outer covering and thus do not need to be ingested to cause infection. Although these qualities make them highly desirable for pest control, many fungi are difficult to produce on a commercial scale and do not persist under field conditions. However, as our knowledge of the genetics, physiology, and growth of fungi increases, these problems might be overcome.

NEW APPROACHES TO ANIMAL AGRICULTURE

Animal Breeding

For centuries, people have sought to improve animal productivity by selecting and breeding only the best animals. Breeders have sought to develop animals that grow bigger, produce more, provide leaner and better quality products, use resources more efficiently, or show increased fecundity or resistance to disease and stress. Compare the average milk yield of dairy cows in the United States today with that of herds 30 years ago: Today half the number of cows are producing the same amount of milk while consuming one-third less feed. This success is mainly the result of controlled breeding efforts, together with improved feeding and other management practices.

Increased understanding of reproductive biology and the genetic basis of traits has given breeders new tools to accomplish these goals. Artificial insemination has revolutionized animal breeding. Embryo transfer for livestock animals is another industry that has changed the nature of cattle breeding in pure-bred herds and has also become important for livestock export. The next important advances in animal agriculture will result from combining conventional breeding methods with new biotechnologies, including genetic engineering. These new methods will give breeders unparalleled precision in manipulating desired traits, and at the same time, they will speed up the process. In the long-term, they may open the door to interspecies gene transfers.

Some applications of biotechnology, such as using monoclonal antibodies as diagnostic aids, have already occurred in animal agriculture. However, the technology of gene transfer in animals is still in its infancy, despite some notable laboratory successes. Molecular gene transfer into animal cells predates similar experimentation with plants. Unlike plants, however, animals cannot be regenerated asexually. Thus, the only way to introduce a foreign gene into all the cells of an animal, including the cells that allow it to pass the trait to its offspring, is to insert the foreign DNA into germ cells—the sperm or the egg—or into the product of their union—the zygote. Another complicating factor is that many production traits—for example, muscle growth, number of offspring, and milk production—are thought to be polygenic traits, meaning they are controlled by the interaction of many different genes. The following sections describe some existing and expected developments in biotechnology that will benefit animal agriculture.

PREGNANCY TESTS

Scientists have developed and patented a monoclonal antibody test to diagnose pregnancy in cows, which could be an important advance for dairy farmers and cattle breeders. The test identifies a protein from cells in the placenta; it can detect pregnancy 24 days after breeding, an improvement over traditional methods. Thus, the farmer can be assured that the cow is pregnant, ensuring the highest efficiency in reproduction. The new test is also more reliable, does not require special skills to conduct, can be conducted on the farm, and gives a simple "yes/no" result similar to the human pregnancy tests now marketed. The test could also benefit zoos and wildlife management specialists, because it is accurate in any ruminant animal, including wild and domestic sheep and goats, elk, deer, and musk-oxen.

In a related development, a British company has developed a monoclonal antibody test that indicates when dairy cows come into estrus. This kit is expected to be marketed soon. An accurate knowledge of estrus is important for the timing of artificial insemination and maintaining maximum milk production. This test, too, can be conducted on the farm.

GROWTH HORMONES

Research efforts that could lead to potentially valuable applications of biotechnology in animal agriculture involve the low-cost production of large quantities of animal growth hormones. For example, bovine growth hormone (BGH) is a naturally occurring hormone that increases milk production in cows. Scientists have been able to genetically engineer bacteria to produce the hormone, which when administered to lactating cows daily can increase milk production up to 40 percent. The animal's milk composition does not change, although it does require greater amounts of and more nutritious feed. However, daily injections of BGH may be impractical for most dairy herds, so researchers are developing injectable slow-release formulations, as well as studying ways to transfer the BGH gene into the animals. The latter process is complicated by the fact that farmers would not want unregulated release of the hormone; they want the gene to be expressed only during lactation to obtain increased milk production.

Porcine growth hormone (PGH) has also been cloned in bacteria, purified, and administered to pigs by injection. PGH greatly stimulates the pigs' growth performance, elevating their growth rate, feed efficiency, and ratio of muscle to fat. These improvements appear to stem from PGH's ability to depress the growth of fatty tissue. Thus, nutrients are redirected to muscle growth. Because PGH is a naturally occurring protein hormone, it is metabolized by the animal. Furthermore, any unmetabolized hormone is broken down during digestion and therefore poses no residue problem to consumers. The impetus for research and development of PGH comes from consumer demand for leaner and therefore more nutritious meat. As with BGH, PGH must be produced at low cost and be easily administered for a controlled delivery over a sustained period of the animal's life. Intense research promises to yield commercial products within 5 years.

Preliminary experiments with mice show that it is possible to regulate gene expression artificially. Scientists have transferred a combination of the growth hormone gene and a segment of DNA that recognizes another group of hormones—the glucocorticosteroids—to see if they can create an "on/off" switch for the gene. Results suggest that feeding mice food that contains these steroids can cause their inserted genes to "turn on" and produce the growth hormone. There is still more to be understood, though, before the technique can be used in livestock and other farm animals. Experiments with animals other than laboratory mice have met with limited success, and have shown side effects such as sterility. However, research on growth hormones and their expression could offer great benefits for a variety of meat animals, including cattle, hogs, poultry, and fish.

BOOROOLA GENE

Gene mapping is essential as the foundation for genetic manipulation. Thus far, however, few specific genes of significance to animal agriculture have been identified, isolated, or mapped. One example of a gene that is beginning to be understood, although it has not been specifically isolated, is the booroola gene from Australian merino sheep. This gene boosts the incidence of twinning and triplets in sheep, giving an overall 20–40 percent increase in the number of lambs weaned. Introducing the booroola gene into other sheep and cattle could offer a fast, reliable way to increase the productivity of ewe and cow herds. Although the gene could be crossed into some breeds by sexual breeding, its introduction by molecular gene transfer would be faster and, more important, it would allow the trait to be passed to a wider range of livestock. Mapping of the booroola gene is helping scientists determine more precisely how the gene operates and is also aiding in its cloning. Scientists may then attempt to transfer the gene to other valuable livestock species.

MUSCLE VERSUS FAT

One goal of animal breeding has been to develop better quality products, such as animals with less fat and leaner meat. The same goal is being pursued by genetic engineers. Working toward this end, biochemists have developed a serum containing antibodies that attack and destroy body fat. Basically, the antibodies bind to specific sites on fat cells; then the animal's natural defense system attacks and destroys the fat. What happens to the dead fat cells is not yet understood, though the degraded free fatty acids appear to be returned to the bloodstream and provide energy to build other body cells. Theoretically, the technique could be applied to any species—pigs, poultry, sheep, or cattle.

FISH FARMING

Despite a long history of reliance on fish as an important source of food, particularly protein, the science of aquaculture is relatively young. Thus, our understanding of genetics, breeding, and reproduction in fish lags behind other agricultural sciences. Tremendous potential exists, however, to use modern technologies, including biotechnologies, to improve aquaculture. One advantage to working with fish is that, in most cases, each fertilization and subsequent development can easily be manipulated. It is possible, for instance, to manipulate the number of chromosome sets in fish eggs to get triploid and tetraploid fish. This technique produces sterile progeny, which helps ensure maximum growth because no energy is "wasted" on reproduction. Scientists can also regulate the sex of fish through various treatments, an advantage because female fish are preferred for commercial markets.

Microinjection of growth hormones is another technique that has been proven effective in promoting fish growth, and genetic engineering of fish to augment their growth hormones is under way. Scientists are also studying ways to genetically engineer fish to be more tolerant of cold temperatures. If an "antifreeze" gene from winter flounder—a cold-tolerance gene present in all Antarctic fish—could be successfully transferred, more types of fish could live at colder temperatures, both for wild propagation and in aquaculture ponds. A number of basic studies of fish molecular biology are under way to increase our understanding of how fish respond to their environment at the molecular level and to develop ways to use this knowledge to increase the efficiency of fish production.

Microorganisms Associated with Animals

Each year livestock and poultry diseases cause an estimated $14 billion in losses. Thus, one important use for biotechnology in animal agriculture will be in the diagnosis, prevention, and control of animal diseases.

Monoclonal antibodies in particular offer great potential for helping scientists understand animal disease. They can be used to diagnose disease, monitor the efficacy of drugs, and develop therapeutic treatments and vaccines to immunize against certain diseases. Monoclonal antibodies are available as therapeutic treatments against both calf and pig scours, which cause at least $50 million in losses annually. Diagnostic tests for these and other diseases—for instance, bluetongue, equine infectious anemia, and bovine leukosis virus—are also already on the market. Applications of such diagnostic and therapeutic products, however, may be limited to high-value animals. Even though the costs involved are not great, farmers work within tight economic constraints and generally cannot afford to routinely use such products.

VACCINES AGAINST ANIMAL DISEASE

Antibiotics are generally ineffective in treating diseases caused by viruses, and many viral diseases go unchecked because there is no appropriate vaccine. Using the tools provided by biotechnology, researchers are working to develop vaccines for many important animal diseases. As mentioned earlier, therapeutic treatments against scours have been developed using monoclonal antibodies. Preventive vaccines have also been developed. These vaccines depend on cloned genes of the disease agent that are used to produce large quantities of certain proteins in cell culture. When injected into animals as vaccines, these proteins stimulate the animal's own immune system to protect it from infection. Foot-and-mouth disease, which affects livestock throughout South America, Africa, and the Far East, is currently a prime candidate for a genetically engineered vaccine.

Such vaccines, derived by techniques of genetic engineering, can be effective, safe, easy to manufacture, and economical to produce. They have long shelf lives, are stable at ambient temperatures, and do not contain lethal infectious viruses—thus avoiding the potential problem of inadvertently causing the disease one is vaccinating to prevent. Genes have been cloned for the surface proteins of viruses that cause fowl plague, influenza, vesicular stomatitis, herpes simplex, foot-and-mouth disease, and rabies, and experiments are leading to the development of vaccines for these animal diseases.

Many questions remain to be answered, however, before routine and widespread use of such vaccines can occur. For the vaccines that are currently being developed, questions remain about side effects, dosage, and timing of vaccination. In addition, some animal diseases of considerable economic significance—such as mastitis—will require extensive basic research before a vaccine can be designed. Research and development of genetically engineered vaccines is time consuming, because each disease, and the many pathogenic strains causing it, must be investigated individually. For each disease, a specific immunogenic antigen must be identified, and the appropriate gene must be isolated and transferred into a bacterium or other fermentable organism such as yeast to allow its manufacture in large quantities.

The first commercial application of a genetically altered vaccine—and actually the first environmental release of an engineered product—is Omnivac, a vaccine that immunizes swine against pseudorabies. Pseudorabies is a serious livestock disease, infecting about 10 percent of the nation's 4 million swine and costing the pork industry as much as $60 million a year. Like earlier vaccines against pseudorabies, Omnivac consists of pseudorabies viruses that are altered to prevent them from causing disease but that are still capable of triggering the production of antibodies. The difference between this and previous pseudorabies vaccines, however, is that Omnivac viruses were altered by genetic engineering—a piece of genetic information was deliberately deleted to incapacitate the virus. Traditional vaccines use imprecise techniques to weaken viruses and pose some, albeit small, danger of causing the disease they are supposed to prevent. Although a controversy arose over the regulatory mechanisms used to approve the Omnivac vaccine, neither proponents nor opponents have questioned the increased efficacy and safety of the product.

VACCINES FROM VACCINIA VIRUS

A nonlethal virus called cowpox was used in the eighteenth century to combat the lethal human disease smallpox. Cowpox was thus the world's first effective vaccine. Scientists subsequently developed the related vaccinia virus into the modern vaccine that eliminated smallpox from the world. Vaccinia is a nonlethal, nonpathogenic virus that conveys a strong and lasting immunity, is easily and cheaply manufactured, and can be transported without refrigeration or loss of potency. Further, it can be injected under nonsterile conditions with a jet gun, a factor that contributed to its success in mass vaccination programs in developing countries.

These and other properties make it an ideal candidate to be genetically engineered to combat other diseases, both of humans and of agriculturally important animals.

Vaccinia is basically a delivery system: Given appropriate protocols, any gene can be moved into vaccinia and be carried into the recipient of the vaccine. This ability means the virus can be adapted to combat essentially any selected disease. Extensive work is necessary, however, to identify, isolate, and transfer the appropriate genetic material. So far, many foreign genes have been inserted and found to be active in vaccinia virus. Vaccinia is a large, complex virus that can simultaneously accommodate at least a dozen foreign genes and still successfully infect cells and replicate. Thus, a single vaccinia vaccine could immunize the recipient against a dozen different diseases. Researchers might someday develop "cassettes," carrying genes for various antigens of the primary infectious diseases in a given geographic area—one for Africa, South America, and so on. A single inoculation would confer immunity to the collection of diseases whose antigenic genes were packaged into the vaccine.

Recombinant vaccinia virus vaccines are more efficient than conventional subunit vaccines that consist of only antigenic protein. The difference is that vaccinia places the genes coding for the pathogen's antigen into the recipient's cells. Antigenic protein is then produced within the cells themselves. This method stimulates the vaccinated recipient's immune defenses more effectively than subunit vaccines, and immunity is longer lasting. Researchers have constructed vaccinia vaccines against a number of human diseases, including hepatitis B, herpes simplex, influenza, and malaria, and against some lethal animal diseases, including rabies and vesicular stomatitis virus. Extensive testing is under way. Animal agriculture will further benefit as scientists develop vaccines against other specific animal diseases.

ALTERING INTESTINAL ORGANISMS

A more speculative area of interest for genetic engineers lies inside agricultural animals. Given appropriate research, could a way be found to alter the intestinal bacteria of ruminant farm animals to make them more efficient in utilizing plant waste fibers for food?

Scientists are looking for ways to improve the microorganisms inside an animal to create a more effective, natural, bioprocessing system. Application of biotechnology to this area is just beginning, but it provides a glimpse of the far-reaching possibilities that lie ahead for agriculture.

BIOPROCESSING OPPORTUNITIES

Several familiar age-old procedures are forms of bioprocessing—fermenting grape juice or leavening bread dough, for example. Yet bioprocessing also includes a range of technologies in which living cells or their components, such as enzymes, are used to cause the desired physical and chemical changes.

Bioprocessing to produce industrial chemicals began during World War I when researchers developed alternative ways to produce acetone and butanol using microorganisms. However, the growth of the petrochemical industry during World War II replaced the microbial production of industrial solvents, and industrial bioprocessing for bulk chemicals practically disappeared. The climate changed again, however, when it was discovered how well biological processes could synthesize complex molecules such as antibiotics, vitamins, and enzymes. The industry was transformed from one that produced high-volume, low-value industrial chemicals to one that produced lower-volume, high-value products.

Advances in biotechnology have renewed interest in industrial uses of agricultural and forestry commodities. Bioprocessing offers innovative opportunities to create new products and foods, treat and use wastes, and use renewable resources (biomass) for fuel. Once developed, such processes could prove more economical as well as less environmentally damaging than current industrial processes.

Alternative Fuels

Many people have hoped bioprocessing could have a significant impact on fuel production, but the present economic situation favors the extraction of natural reserves of petroleum, gas, and coal. Biomass energy, such as alcohol produced from grains and sugar, or methane (biogas) produced from animal manures and other waste products, has received some research attention. In the United States, gasohol (consisting of 10 percent alcohol and

90 percent gasoline) made a brief, well-publicized appearance, but price changes in the oil market have undermined its competitiveness. In Brazil, alcohol fuel is widely used; it is obtained primarily from the fermentation of sugarcane juice.

However, producing energy from food crops is not yet profitable in most countries. Most of the sugar- and starch-containing plants, such as potatoes, corn, and cassava, that are easily converted into alcohol are relatively expensive. In addition, widespread and large-scale use of food crops for energy production could create food shortages, especially in developing countries.

However, as scientists engineer microorganisms to feed on cellulose and develop efficient ways to break down the lignin (the tough compound that makes wood resistant to degradation) in woody plants, a fuel-alcohol industry based on less valuable plant materials (including trees, weeds, scrub, and wastes from paper manufacturing) might be developed. Similarly, the potential for bioprocessing to create methane lies in using microbes and wastes—domestic sewage, manure, crop residues, and other cheap and available raw materials. Some scientists foresee a time when bioprocessing might also be developed to produce hydrogen for fuel.

Progress in developing bioprocessing for alternative fuels will occur slowly because vehicles and markets adapted for such fuels are not developed, and there are no economic incentives for these markets to change. In addition, bioprocessing for bulk chemicals or for energy (e.g., methane, methanol, ethanol, etc.) is difficult to engineer even with a uniform feed stock such as sugarcane or corn. When a diversity of biomass materials is used, problems are compounded by the design of fermentation apparatus and the selection of microorganisms adapted to grow on different feedstocks. Continued research on bioprocessing for bulk chemicals and alternative fuels, however, is important. Opportunities to use inexpensive by-products or wastes, or changes in economics based on the price of oil and gas, may make it economical in the long-term.

Alternative Feed and Food Sources

Bioprocessing also holds promise as a way to create unique sources of protein for an increasingly hungry world. For instance, scientists have found some unusually hardy microbes living in the Dead Sea, and one of these, *Dunaliella bardawil*, manufactures glycerol to counteract the pressures of its highly saline environment. In Israel, this alga is grown and harvested in specially built ponds. In addition to glycerol, manufacturers obtain a compound called beta-carotene that is sold as a food coloring and a residue that is an excellent, protein-rich animal feed. In Finland, sulphite liquor from paper production is fed to certain molds, which not only purify the waste liquids but also yield a rich residue that is sold as animal feed. Similar techniques could be developed for waste materials from forestry, cheese-making, and other industries.

Microbes have long played a role in food production. Cheese, pickles, bread, beer, and wine, for instance, all rely on bioprocessing. Molecular genetic techniques are being used to monitor the properties of microbes used in these processes to ensure product uniformity. Yet microbes can do more than preserve foods or alter their taste; the future might include a direct microbe-based food source: single-cell protein. People have consumed microbes in the form of algae as far back as the Aztecs. Modern biotechnology looks to single-cell protein primarily as an animal feed, but some scientists consider human consumption a possibility, too.

Other Products

Bioprocessing already contributes to our ability to produce vitamins, amino acids, enzymes, and more recently hormones, and this role should increase in the future. For instance, much of the supply of vitamins B_2 (riboflavin) and B_{12} (cobalbumin) comes from microbes. Researchers have adapted wild strains of a mold, *Ashbya gossypii*, to produce 20,000 times its original output of vitamin B_2. Research has also intensified microbial production of vitamin B_{12} over 50,000 times.

Most cereal grains are deficient in two essential amino acids, lysine and methionine. These are usually added to animal feed to ensure an adequate diet. Methionine is made by chemical processes, but 80 percent of all lysine is produced by fermentation using bacteria. The amino acid derivative monosodium glutamate, which is used as a flavor enhancer in cooking, is produced by two bacteria through a bioprocess.

In the past, the use of enzymes has been limited by the expense of isolating them from natural sources and by their instability. Recent advances have provided ways to immobilize enzymes and use whole microorganisms as catalytic systems, thus yielding more stable and reusable enzymes and increasing the opportunities for their use. The biotechnological production of sugar substitutes is one example of a growing industry that has been made possible largely because of our increased ability to manufacture enzymes through microbial processes.

Another area with potential for bioprocessing is waste treatment. As mentioned in previous examples, some bioprocessing systems can transform plant debris and other wastes into useful products, in effect creating an inexpensive and abundant renewable resource. Another current example is a new strain of yeast genetically engineered with an enzyme that converts the lactose in whey, a dairy industry waste product, into ethanol, which has fuel energy value. On another front, bioprocesses are being developed to more efficiently treat municipal, industrial, and agricultural wastes. However, some problems remain in improving the dependability and design of these systems.

To develop new approaches toward bioprocessing and to bring them into widespread use will require a great deal of research. First, systems must be designed to accomplish each goal. Successful systems will require (1) a solid understanding of the organism involved, (2) an effort to develop the most productive strain of the organism and isolate the appropriate enzymes, and (3) intensive, specific research on the dynamics of each individual bioprocess. Next, the bioprocessing systems must be improved and perfected to offer economically competitive products. Research is needed to develop new industrial-scale methods to isolate products at the degree of purity appropriate for commerical use. Concentration of the final products is also important because separating them out after microbial conversion is often a major cost.

CONCLUSIONS

Benefits offered by biotechnology will not be fulfilled without a continued commitment to basic research. In fact, most of the prominent new biotechnologies are "spin-offs" from basic research efforts. As the examples in this chapter indicate, improved yields and reproduction, disease resistance, better quality products, reduced inputs, and similar advances are possible using biotechnology. However, society must be prepared to support the long-term efforts needed to transform these ideas into practical applications. Extensive laboratory and field research will be necessary to develop specific applications. This research will require considerable time and funding. Some of these new developments could dramatically transform agriculture and food production by increasing efficiency and productivity, thus lowering costs and improving competitiveness in the world marketplace.

If we are to continue to make progress using genetic engineering to improve agriculture—whether by engineering the plants, animals, or the microorganisms and insects associated with agriculture—research must focus on six important areas.

1. Gene identification—locating and identifying agriculturally important genes and creating chromosome maps.
2. Gene regulation—understanding the mechanisms of regulation and expression of these genes and refining the methods by which they may be genetically engineered.
3. Structure and function of gene products—understanding the structure and function of gene products in metabolism and the development of agriculturally important traits.
4. Cellular techniques—developing and refining techniques for cell culture, cell fusion, regeneration of plants, and other manipulations of plant and animal cells and embryos.
5. Development in organisms and communities—understanding the complex physiological and genetic interactions and associations that occur within an organism and between organisms.
6. Environmental considerations—understanding the behavior and effect of genetically engineered organisms in the environment.

GENE IDENTIFICATION

Gene identification is crucial to the advancement of biotechnology, because scientists need to understand what gene is responsible for the trait they want to alter. Basic research in biochemistry and genetics is necessary to be able to identify specific genes and the traits associated with them. Only after the specific gene is identified can scientists alter it to benefit agriculture. Thus, it is important that our ability to identify genes be improved for future advancements in biotechnology.

Chromosome Maps. Although they are merely general catalogs of a plant, animal, or microbial genome, chromosome maps are important guidelines for finding specific genes of importance to agriculture. Chromosome maps can show genetic engineers where to begin their search for specific genetic information. Chromosome maps identify "markers" that are often linked to important genes, such as the gene for a specific disease or physical trait, and they can be used to trace inheritance patterns. In humans, we have learned what markers, rather than specific genes, are linked to some inherited diseases such as cystic fibrosis. Researchers could provide a powerful tool to aid in the development of biotechnology if they would develop chromosome maps for the major crop species such as corn, wheat, and rice, and for important animal species such as cattle, swine, and poultry.

GENE REGULATION

Once a gene has been identified, the importance of understanding gene regulation becomes clear. Part of manipulating a gene is getting it to be expressed appropriately. To accomplish that, scientists must understand how the gene is controlled—what turns it on and off, how it interacts with various hormones, and other factors. The science behind gene regulation is very intricate and requires a sophisticated understanding of molecular biology. Gene regulation becomes especially complex when several genes interact to control a trait. Such "multigenic" control is involved in some important agricultural traits, for instance in determining the storage proteins that contribute to the nutritional value of a crop or its hardiness in a particular environment. Advancing our understanding of gene regulation and expression will require basic research in biochemistry, physiology, and genetics, and will require intensive laboratory research, because each gene must be studied as an individual case.

STRUCTURE AND FUNCTION OF GENE PRODUCTS

The end products of the actions of genes are of prime interest in agriculture. The cellulose fibers of trees and cotton, the proteins in seeds or muscle fibers, and the carbohydrates and fats important in food and commerce are the end products of highly organized and regulated metabolic pathways. Genes code for the enzymes as well as for the structural and regulatory molecules that carry out the complex reactions that lead to these end products. The deficiencies of our understanding in the biochemistry and physiology of metabolism and development are often the greatest constraints to applying biotechnology to agriculture. Understanding the linkage between metabolism and development and the genes that encode these processes will require progress on both fronts. The tools of biotechnology and techniques for isolating and manipulating genes can aid biochemical and physiological studies of metabolism. Conversely, studies of metabolic pathways can help us identify genes and understand their regulation.

CELLULAR TECHNIQUES

The manipulation of plant and animal cells is part and parcel of strategies that involve genetic engineering, monoclonal antibodies, and bioprocessing. Although methods for cell culture, cell fusion, regeneration of plants from cells, and embryo manipulation exist for some species, these techniques must yet be successfully adapted to other species, which include important crops and livestock animals. Moreover, specific microorganisms such as yeasts, fungi, viruses, and bacteria important to agriculture and bioprocessing must be able to be cultured to allow both basic research and practical applications.

DEVELOPMENT IN ORGANISMS AND COMMUNITIES

Genetic engineering is more complex when it involves interactions among organisms. The symbiotic relationship between a microorganism and its host plant is intricate and raises many questions for scientists. Gene identification remains important: What genes are involved in various stages of the relationship? Why does the microorganism colonize only one type of plant? Detailed study is necessary to answer these sorts of questions about particular relationships.

Researchers also need to understand the relationships under field conditions if they are to design organisms that can compete effectively once they are released. Another aspect of the associations between plants and microorganisms that needs research involves the mechanisms of infection. Knowing how a microorganism attacks a plant is the first step in combating it. Without that basic understanding, genetic engineers will not be able to manipulate the system to their advantage. Although genetic manipulation is becoming a reality, in far too many cases a lack of understanding of plant physiology and pathogen interactions limits its progress.

ENVIRONMENTAL CONSIDERATIONS

Many of the pending applications of biotechnology will require releasing genetically engineered plants, animals, and microbes into the environment. Clearly, the more that is known about the ecology and behavior of plants, animals, and microorganisms, the better are our chances of assessing the potential values and possible risks involved in introducing genetically altered versions into the field. Data on pathogenicity, mutagenicity, the ability to transfer genes, and other relevant factors can help predict the organism's effects on the ecosystem. Indeed, developing data and tools to support value and risk assessment is likely to become an increasingly important part of research efforts. The results of such work will help scientists understand the system, and will play a role in educating the public about both the risks and benefits offered by biotechnology. However, a detailed analysis of the regulatory aspects of this important and controversial issue is beyond the scope of this report.

RECOMMENDATIONS

INCREASED EMPHASIS ON BASIC RESEARCH

Basic research programs in physiology, biochemistry, genetics, and molecular biology within agricultural disciplines such as agronomy, entomology, and animal science need to be strengthened and in many cases redirected to questions of identifying genes and understanding the regulation of their expression. Just understanding the regulation of their expression. Just as an enormous information base has provided a substructure for sweeping advances in biomedical science, a similar foundation of knowledge is now needed about the basic biochemistry, physiology, and genetics of such agricultural subjects as host–pathogen interactions, plant and animal developmental responses to environmental stimuli, enzymes and metabolic pathways, and molecular constituents and their patterns of organization in subcellular organelles. Acquiring such knowledge will affect the rate at which agriculturally valuable genes can be identified, isolated, and characterized, and is a prerequisite for applying the tools of biotechnology to agricultural problems.

A similar call for augmented basic research within agricultural and related biological and biochemical fields was sounded in previous reports (NRC, 1984, 1985a, 1985b; Winrock International, 1982). Positive steps have been taken. Yet far more impetus is needed to ensure the continued success of American agriculture in an ever-changing world economy.

IMPROVED TECHNIQUES AND APPLICATIONS

The repertoire of molecular biology and cell culture techniques needed to implement advances in genetic engineering is incomplete. Methods for gene transfer in many plants, animals, and microbes; plant cell culture and regeneration; and animal embryo culture and manipulation are inadequate to support the goal of improving agricultural productivity. Increased efforts are needed to apply techniques developed for laboratory organisms to those plants, animals (including insects), and microbes relevant to agriculture.

A national effort should be mounted by both public and private sectors to apply techniques of biotechnology to problems in the agricultural sciences. This effort should include research on:

- Gene identification—locating and identifying agriculturally important genes and creating chromosome maps.
- Gene regulation—understanding the regulation and expression of these genes and refining methods by which they may be genetically engineered.
- Structure and function of gene products—studying the structure and function of gene products in metabolism and the development of agriculturally important traits.

- Cellular techniques—developing and refining techniques for cell culture, cell fusion, regeneration of plants, and other manipulations of plant and animal cells and embryos.
- Development in organisms—using the new technology to study cell and organismic biology in intact organisms.
- Development in communities—understanding the complex associations and interactions that occur among organisms.

INCREASED ATTENTION TO THE ECOLOGICAL ASPECTS OF BIOTECHNOLOGY

Both the public and private sectors should increase their efforts to develop an extensive body of knowledge of the ecological aspects of biotechnology in agriculture. In particular, studies must be done to further our understanding of the behavior and effects of genetically engineered organisms. In addition, the public must be educated about biotechnology. These efforts are essential to support future applications of biotechnology and to adequately inform regulators and the public about both the benefits and possible risks involved.

3
Funding and Institutions

FUNDING BIOTECHNOLOGY IN THE AGRICULTURAL RESEARCH SYSTEM

Any strategy to use the tools of biotechnology to advance agriculture and forestry must address funding and the institutions of the research system. Funding and institutions are the foundation for progress in biotechnology. These two factors nurture and shape the development of new knowledge, the training of scientists, and the implementation of technical innovations. As tools of biotechnology are adapted to the problems of agriculture, new demands will be placed on the existing arrangement of research institutions. Similarly, biotechnology also will influence patterns of funding for research and training and may alter the established pathways between research discoveries and applications. The pace at which biotechnology is applied to agriculture depends on how rapidly the R&D system can incorporate these changes.

This chapter looks at current institutions and funding patterns in agricultural research and how they are changing with the advent of biotechnology. It examines ways to enhance the roles of the federal government, states, and private sector in supporting biotechnology research. It also calls for greater use of peer-reviewed, competitive grants to guide the growth of the agricultural biotechnology research system. In addition, it calls for greater integration of basic and applied research.

The Federal–State Agricultural Partnership

The USDA and the land-grant university system, both created in 1862, have long been the keystones of our national agricultural research system. This decentralized system creates close ties between federal and state programs and farmers. Up to now, the enormous success of U.S. agriculture has been credited to the strength and character of this network, especially its abilities to solve important problems and coordinate agricultural research and extension services at the federal, state, and local levels.

The federal institution chiefly responsible for agricultural research is USDA, which supports research and extension through the Agricultural Research Service (ARS), the Cooperative State Research Service (CSRS), the Forest Service, and the Cooperative Extension Service (CES). ARS and the Forest Service are primarily the in-house research agencies of the department; CSRS and the CES direct and coordinate federal funds and special grant programs to the states. At the state level, the land-grant colleges of 1862 and 1890 and Tuskegee Institute support research, training, and extension programs in agriculture. The State Agricultural Experiment Stations (SAESs) and the State Cooperative Extension Services, which are partly supported by federal appropriations, are attached and integrated (with a few exceptions) into the land-grant universities. Many county governments also are involved in agricultural extension, but their level of financial support and role in extension activities varies within as well as between states.

Federal appropriations to the states for research and extension programs require approval by CSRS or CES. This arrangement of co-funding by states and the federal government provides an avenue of input from both sides in the partnership. It is the basis of a nationally coordinated yet decentralized research and extension system in agriculture.

It is not easy to characterize the workings of the many priority-setting mechanisms and processes determining the direction of the research and extension system. In the federal–state partnership for supporting agricultural research, state and local concerns have tended to predominate. This is not surprising because most people in the system are state and not federal employees. However, the federal budget-making process has a major impact on the financial resources available.

A number of organizations act to coordinate planning and set priorities in the research and extension system. Within the Division of Agriculture of the National Association of State Universities and Land-Grant Colleges (NASULGC), there is the Experiment Station Committee on Organization and Policy (ESCOP) and the Extension Committee on Organization and Policy (ECOP). At the federal level, the 1977 farm bill established a Joint Council on Food and Agricultural Sciences and a Users Advisory Board to advise Congress and the Secretary of Agriculture. The membership of these two advisory groups includes representatives from private companies, foundations, and non-land-grant universities, as well as the traditional federal and state agricultural agencies.

Finally, the system includes federal and state legislative committees and executive institutions that may influence or have budget control over public agricultural research programs and policy. Also involved indirectly are the General Accounting Office (GAO), the Office of Technology Assessment (OTA), and within the Executive Office of the President, the Office of Management and Budget (OMB) and the Office of Science and Technology Policy (OSTP).

Past Contributions from Agricultural Research

Historically, agriculture has relied on public investment in both basic and applied research. This reliance is particularly true for certain research areas such as cultural practices or fundamental breeding programs, in which the private sector cannot easily create a "product" and thus recoup its investment. Studies have demonstrated that public investment in agricultural research produces a very high rate of return. Research expenditures worldwide provide annual rates of return of about 50 percent (Evenson et al., 1979).

During some periods the rates of return in American agriculture have been even higher. For example, from 1927 to 1950 the returns of technology-oriented research in agriculture were estimated to be 95 percent. The returns of science-oriented research were even higher—110 percent. Technology-oriented research was defined as including such areas as plant breeding, agronomy, animal production, engineering, and farm management. Science-oriented research included soil science, botany, zoology, genetics,

plant pathology, and plant and animal physiology. The higher return from science-oriented research is noteworthy considering that
biotechnology relies on a new array of disciplines oriented toward
basic science.

Research has contributed to increased agricultural productivity, low and stable food prices for American consumers, and
enhanced competitiveness in world markets. Much of this past
success can be attributed to the "articulation" and "decentralization" of the American agricultural research establishment (Ruttan, 1982). The close links among various parts of the system—
basic research, applied research, extension, private industry, and
farmers—were strengthened by the decentralization of authority
to the state and local level. Yet it has also been argued that in this
decentralized research system, basic research has been underval-
ued and underfunded. Some even suggest that this underfunding
of basic research explains, in part, the high rates of return. For
example, spillover of basic biomedical research discoveries benefits
agricultural research, but the costs of such biomedical research are
not factored into rate of return estimates for agricultural research.
Overall, however, the continuous state and federal support for research in the land-grant college system has benefited American
agriculture and society at large for close to a century.

Pressures for Change

Despite the past successes of the nation's agricultural research
and extension system, it is not without its critics and problems.
By the early 1970s there were signs that the unique approach of
the federal–state–community alliance had in an unforeseen way
separated agriculture from the rest of academic science. One analysis concluded that agriculture—"the mother of sciences"—was
an island empire, isolated from American academic life and no
longer at the leading edge of scientific progress (Mayer and Mayer,
1974). Another analysis, known as the Pound Report, argued
that public agricultural research had become highly insular and
divorced from the frontiers of knowledge in the basic biological
sciences (NRC, 1972). This report and others that followed recommended strengthening support for the basic plant and animal
sciences (Brown et al., 1975; NRC, 1975; OTA, 1977, 1981; Winrock International, 1982). These reports urged the agricultural

research system to establish new funding programs based on open competition with scientific merit determined by a process of peer review—the same process used by other federal agencies to award research grants in the sciences.

Even as the enormous successes of the "Green Revolution" were introduced into developing nations around the world, critics of the agricultural research and extension system were pointing to problems the system had failed to address. In her famous book *Silent Spring* (1962), Rachel Carson called public attention to environmental issues and the problems created by the widespread use of pesticides. Agricultural research that had deciphered the interactions of soil, water supply, climate, and pests in crop production now needed to address broader environmental and ecological problems. The agricultural research system also was criticized on the grounds of social equity and social justice. *Hard Tomatoes, Hard Times* argued that the land-grant college system, initially established to serve the mass of rural and agricultural people, had become a publicly subsidized research arm serving agribusiness and the large farmer (Hightower, 1973).

Although buffeted by criticism and increasing public demand to broaden agriculture's research responsibilities and to encompass scientists from allied disciplines, few dramatic changes in either the institutions themselves or in funding patterns have been implemented. The National Agricultural Research, Extension and Teaching Act of 1977, which is Title XIV of the Food and Agriculture Act of 1977 (P.L. 95-113), did authorize a series of new research and education grants and fellowships. One of these was a program to support high-priority research through a competitive grants program available to SAESs, all colleges and universities, other research institutions and organizations, federal agencies, private organizations or corporations, and individuals. Authorization was made for appropriations up to $25 million for the program in 1978 with $5 million increases in the subsequent 3 years and a $10 million increase for 1982, for a total of $50 million. However, actual appropriations made by Congress fell far short: only $15, $15, $15.5, $16, and $16 million were appropriated for those 5 years.

This lack of commitment to financial support for basic research in agriculture has had cumulative and far-reaching impacts: "Congress has held research resources constant for 15 years and since World War II has slowly politicized and destroyed the magnificent science investment in the old UDSA biological and physical science bureaus and the successor Agricultural Research Service" (Bonnen, 1983).

Demands and pressures on the federal–state partnership and on the research system as a whole remain. Yet support from the federal government has not been sufficient to accommodate these growing needs. At present the opportunities and needs of biotechnology in agriculture are being added on top of existing demands and pressures. In 1983 a report from the Division of Agriculture of the National Association of State Universities and Land-Grant Colleges (NASULGC, 1983) called for increased funding by the federal government of at least $70 million per year in competitively awarded grants to support research and education programs in biotechnology related to agriculture. The report also stated that even a $70 million per year increase would provide funding assistance for only a small portion of the biotechnology programs needed to augment current agricultural research. Congress responded to this and other recommendations for increased support with appropriations in FY85 and FY86 of $20 million to increase the competitive grants program in agricultural biotechnology. The federal government has not responded fully to the call for an increased financial commitment for basic agricultural research.

The Emergence of Biotechnology

The emergence of biotechnology has stimulated and strengthened the contributions of the basic science disciplines of molecular biology and molecular genetics to the agricultural research establishment. It has also placed a stronger emphasis on basic research in cell biology, physiology, and biochemistry. A complete analysis and understanding of the structure, function, and regulation of a gene is usually needed before it can be used for a specific purpose. Such analysis requires a substantial investment of time, talent, and funds before practical applications can be devised.

The types of products that can be developed using biotechnology depend on earlier investments made in basic research. For example, scientists spent years isolating, purifying, and characterizing the coat proteins of the foot-and-mouth disease virus.

However, once they had the amino acid sequences of these proteins in hand, it took them only a few months' work with the tools of biotechnology to prepare subunit vaccines that protect against this costly cattle disease. Similar progress against other diseases will depend on obtaining basic knowledge of the disease agents involved. The development of genetically engineered animal growth hormones and plant herbicide-resistance traits were possible because of the years of fundamental research invested in trying to understand the basic biology of these systems.

In addition to requiring a large initial investment to acquire basic knowledge, biotechnology research approaches shorten the time between discovery and technology development. This is bringing about a greater confluence of basic and applied research interests.

Tools of biotechnology are rooted in discoveries from basic research investigations conducted by the biomedical research community. Although agriculture is predicted to be a major beneficiary of the advances brought about by biotechnology (OTA, 1983), the agricultural research system provided very little support for early developments in biotechnology. Most of the support for research that established the theories and methodologies of biotechnology came from the National Institutes of Health (NIH) and the National Science Foundation (NSF), predominantly in the form of peer-reviewed, competitive grants. Furthermore, most of this research was conducted in private and public university departments with little or no direct connection to the agricultural sector.

Table 3-1 shows levels of support to universities for basic, applied, and developmental research by the major federal research-supporting agencies. Although it is often difficult to make sharp distinctions among these three categories of research, the data show that, except for the Department of Defense, the USDA gives the least emphasis to basic research.

A distinguishing feature of biotechnology is that its unique genetic products are often patentable. Prior to 1970, private sector agricultural research in the United States placed relatively little emphasis on developing biological inputs, with the exception of hybrid seeds, and focused instead on machinery and chemical inputs. However, the Plant Variety Protection Act of 1970 and a 1980 U.S. Supreme Court decision (*Diamond* v. *Chakrabarty*)

TABLE 3-1   Expenditures in FY85 for R&D at Universities by Major Federal Agencies (millions of dollars)[a]

Major Support Agencies	Basic Research	Applied Research	Development	Total	Percentage of Basic Research[b]
DOD	408.8	178.4	352.8	940.0	43
DOE	211.3	124.6	21.6	357.5	59
HHS	2,091.4	889.2	166.9	3,147.5	66
NASA	176.9	36.5	41.6	255.0	69
NSF	943.1	58.7	0.0	1,001.8	94
USDA	142.6	149.6	1.0	293.2	49
Total funding	3,974.1	1,437.0	583.9	5,995.0	

NOTE: DOD = Dept. of Defense; DOE = Dept. of Energy; HHS = Dept. of Health and Human Services; NASA = National Aeronautics and Space Administration; NSF = National Science Foundation; and USDA = U.S. Dept. of Agriculture.

[a] Estimates reflect each agency's classification system and definition of basic and applied research.

[b] Basic research calculated as a percentage of total estimated support. Values are rounded to the nearest whole number.

SOURCE: Federal Funds for Research and Development: Fiscal Years 1985, 1986, and 1987, Volume XXXV, Detailed Statistical Tables. National Science Foundation. Washington, D.C.

established the legality of obtaining patents for novel life forms. These actions have stimulated private investment in agricultural research, and over the past decade, private sector investment in biotechnology has grown sharply. Yet there have been financial casualties. It is difficult for a small company to survive the long gestation period of basic research needed before a product is developed and profits can be realized. The private sector increasingly recognizes that its own progress in biotechnology development depends on the progress made in publicly supported basic research.

Thus, in biotechnology there appears to be an alliance emerging between public sector basic science and private sector technology development. For the most part, these alliances in biotechnology include new participants who have not been part of the traditional agricultural research establishment. Their work and interests complement rather than replace the traditional, public and private agricultural research establishment.

The major issue facing the application of biotechnology to agricultural problems is how to strengthen and link the new and traditional research elements. Advances in basic biological research and applications of the tools of biotechnology are increasing the demand for both public and private sector applied research aimed at technology development and transfer. Meeting this demand is an urgent but formidable task, and will require a significant investment in training and institutional development for research and technology transfer.

INSTITUTIONS THAT SUPPORT AGRICULTURAL RESEARCH

To examine the type of institutions needed to advance agricultural biotechnology, this section looks at who is conducting and funding agricultural research. It then examines the types of institutional and funding changes needed to apply the tools of biotechnology to agriculture more rapidly.

USDA is the primary federal agency supporting agricultural research, but it is only one element in the nation's research system. Other federal agencies, such as the Department of Energy (DOE), EPA, NIH, NSF, and even the National Aeronautics and Space Administration (NASA) make direct and indirect contributions of varying degrees of importance. In addition, the states and the private sector provide extensive support for agricultural research. Together, this combination of federal, state, and private support has brought about significant progress in agriculture. Applying this same level of investment to biotechnology could revolutionize agriculture.

The following discussion highlights the major institutions that support research related to agriculture and gives some indication of their involvement in biotechnology. For federal agencies, the total FY86 appropriation is given in parentheses. However, many of these agencies have only a minor interest in agriculture, and an even smaller interest in biotechnology, so only a small fraction of their research funds are used for these purposes.

Federal Agencies

U.S. DEPARTMENT OF AGRICULTURE

Agricultural Research Service.   The ARS is the primary intramural research agency of the USDA. It conducts research on a range of topics including soil and water resources, environmental quality, the biology and production of crop plants and animals, pests, nutrition, marketing, and international trade. With an annual budget that just reached half a billion dollars (FY86 appropriation: $509.7 million), the ARS supports a network of 133 research centers located across the United States and abroad. Research programs are generally national in perspective and include high-risk, long-range research as well as applied goals. In addition, the ARS maintains genetic stocks of farm animals and plant collections in clonal and seed repositories.

Biotechnology research represents only a small part of the total agricultural research funded by USDA through ARS. According to data collected by the U.S. General Accounting Office (GAO, 1985), as of October 1984, ARS reported that it was conducting 183 biotechnology research projects with an estimated cost in FY85 of $26.4 million. Data collected the following year put the estimated FY85 expenditure for biotechnology research at $24.5 million (GAO, 1986).

Cooperative State Research Service. The CSRS administers federal funds provided for agricultural research at the SAESs and other eligible institutions (FY86 appropriation: $288.7 million). CSRS also participates in the national system of agricultural research planning and coordination, facilitating cooperation among state institutions as well as between state institutions and their federal research partners. In most states, federal funds account for less than one-third of the SAESs' total operating costs.

More than half of the federal CSRS appropriation is distributed under the Hatch Act (FY86 appropriation: $155.5 million). Hatch funds go to the states based on a formula established by Congress that considers the size of each state's rural and farm populations. The SAESs allocate the money for designated projects according to their own priorities. Federal McIntire-Stennis funds support forestry research at SAESs (FY86 appropriation: $13.0 million). A third category of support to SAESs are Special Grants (FY86 appropriation: $28.6 million), usually awarded by Congress and directed to specific agricultural problems at eligible cooperative institutions.

The CSRS Competitive Research Grants (FY86 appropriation: $42.3 million plus $6.5 million for forestry grants) are peer-reviewed and awarded on a merit basis to competing research

TABLE 3-2   Competitive Grant Funding per Principal Investigator in Agriculture, Biology, and Biomedicine

Sponsoring Agency	Average Grant Award per Year[a] (FY86 Awards)
USDA	$ 46,200[b]
NSF	70,000[c]
DOE: Biological Energy Research Division	72,000
NIH	164,000

[a] Values given include both direct and indirect costs.

[b] Competitive Research Grants Office, Forestry, and Small Business Innovation Research Grants.

[c] Plant biology and biotechnology-related grants; the average grant size over the entire Directorate for Biological, Behavioral, and Social Sciences was $65,000.

SOURCE: Personal communications from agency program directors, 1987.

scientists throughout the U.S. scientific community. Competitive grants are given for research projects in animal and plant biotechnology, pest science, animal science, plant science, human nutrition, and forestry. (The forestry grants are a separate appropriation from the U.S. Forest Service, as will be described.) Funding for the competitive grants program increased from $16.4 million in 1984 to $51.7 million in 1985, but declined to $48.8 million in 1986. Of the 1985 funding, $19.2 million was for a component of the grants program to specifically support biotechnology research. This represented 32 percent of the grants and 37 percent of the program funds awarded. In 1986, $18.0 million was allocated for biotechnology research, which is 36 percent of the program funds awarded. Thus, biotechnology-related research now constitutes a major part of the research supported by this grants program. Competition is keen for competitive research grants; only 19 percent of the proposals submitted in all areas were funded in 1985 and 1986. The average grants awarded in 1985 and 1986 were $102,000 and $92,400, respectively, for 2 years or about $51,000 and $46,200 per year (Table 3-2); these amounts are far short of the level of funding required by a modern laboratory to do top-quality research in any of the fields represented.

Forest Service.   The Forest Service is mandated to be a multiple-use agency responsible for managing national forest lands (FY86 total appropriation: $2.5 billion; $120.1 million of this went to research, of which $6.5 million was administered through the CSRS Competitive Research Grants Program). Timber, grazing, recreation, energy development, wildlife conservation, mining, and fire and atmospheric research are all within the agency's research responsibilities. The agency also conducts some research involving biotechnologies, most of which is funded through the competitive grants program. Areas being investigated include research to use biotechnologies to advance genetics in forestry, such as gene identification and transfer to improve species; research to develop products by genetic engineering, in particular, to develop a microbe to create ligninase, an enzyme that helps digest wood waste; research to speed up screening for resistance to environmental stresses and diseases using somaclonal techniques and efforts to transfer genes to convey resistance to selected herbicides; and research on methods to enhance biological control agents that are essential to integrated pest management strategies.

Economic Research Service.   The ERS is an economic analysis unit within USDA (FY86 appropriation: $46.1 million). ERS conducts economic forecasting, policy analysis, and other social science research, often in conjunction with the CSRS and the SAESs. ERS does not have a division devoted specifically to technical analyses, so it is not possible to calculate how much of its budget is allocated to biotechnology-related research. ERS is following biotechnology developments, however. In particular, it is studying biotechnology innovations and conducting in-depth economic and policy analyses of some of these. ERS plans to do studies that focus on the impacts of biotechnology on agricultural competitiveness, structure, and policy as well as the legal and policy aspects of biotechnology itself.

Cooperative Extension Service.   The CES is a nationwide system of federal, state, and local experts that is the primary mechanism for the delivery of research from the land-grant universities to farmers, ranchers, and others. It can also serve as a feedback mechanism to bring problems occurring at the farm and community level to the attention of university researchers. Major program areas in the CES include agriculture, home economics, community development, and 4-H youth programs. State and local funding to CES (FY86 appropriation: $711 million) is more than double the federal contribution (FY86 appropriation: $342.7 million). States contribute funds through the land-grant universities (FY86 appropriation: $486 million), counties contribute locally or to a land-grant university (FY86 appropriation: $193 million), and additional money comes from private funding to land-grant universities and local user fees (FY86 appropriation: $33 million).

The large-scale use of the products of biotechnology in agriculture is still in the future, and thus far CES has not had to focus efforts on providing information and extension services for biotechnology applications. However, as biotechnology products come to market over the next few years, CES will need to hire and train extension agents able to handle this technology. The role of CES in technology transfer for biotechnology is covered in Chapter 5.

Animal and Plant Health Inspection Service. APHIS is mandated to protect U.S. animal and plant resources from diseases and pests using survey, diagnostic, control, and eradication programs and other regulatory activities (FY86 appropriation: $314.4 million). APHIS's major role is that of a regulatory agency, and as such it will play an important part in the regulation of biotechnology in the areas of animal viruses, veterinary services, and plant pathology. Although APHIS does carry out some applied research (FY86 appropriation: $8.2 million) related to its mandate, it is not a research agency.

The Veterinary Services division does some developmental work in projects dealing with diagnostic tests for animal diseases and testing of vaccines for efficacy, although most of its laboratories do actual testing, not research; these projects are beginning to use the tools of biotechnology. The Plant Protection and Quarantine division does developmental work on detection and control of pest insects and weeds, and it is also implementing biotechnology approaches. The Animal Damage Control division does developmental work on behavioral and chemical methods of dealing with animal damage problems, but thus far it has not made use of biotechnology.

ENVIRONMENTAL PROTECTION AGENCY

The EPA will regulate the products of agricultural biotechnology under two acts: pesticides under the Federal Insecticide, Fungicide, and Rodenticide Act and nonpesticidal microorganisms under the Toxic Substances Control Act. Part of the EPA's mission is also to fund actual research on risk assessment. Approximately one-fourth of the agency's operating budget is devoted to R&D (FY86 R&D appropriation: $324.6 million). Some of this research is relevant to agriculture: EPA allocated $11.8 million in FY86 toward research on the impact of chemical and biological pesticides.

NATIONAL SCIENCE FOUNDATION

The NSF has the primary responsibility in the federal government for fostering a strong national ability to conduct basic research (FY86 appropriation: $1.5 billion). The research it supports is usually performed at colleges and universities and other nonprofit research institutions. Proposals for funding are peer-reviewed by panels of scientists; awards are made on the basis of scientific merit and competitiveness. Funding for research related to agriculture occurs primarily through NSF's Directorate for Biological, Behavioral, and Social Sciences (FY86 appropriation: $248.9 million). Much of the biological research funded by the directorate contributes to the knowledge base required to solve practical problems in the areas of health, energy, the environment, and agriculture. However, the NSF contributes only 15 percent of the total federal support in the biological sciences, whereas NIH provides 75 percent (Intersociety Working Group, 1986). On the other hand, within its 15 percent share of federal support, NSF funds over 50 percent of the total federal research effort in plant biology that is supported by competitive grants.

Since its inception in 1952, NSF has played an important role in funding leading-edge basic research in plant biology. This support is significant for agriculture because fundamental research on plants, unlike similar research on animal systems that can be related to human health and thus funded under the huge NIH umbrella, is largely excluded from NIH funding, and basic plant research has not received appropriate attention from USDA. To illustrate, total federal support in 1985 for research on plants awarded through competitive grants was about $110 million, $55 million of which came from the NSF. However, this total was only 5 percent of the $2.2 billion awarded for all federal competitive grants for basic research in biology and biomedicine (NSF, 1986). USDA should logically support research on plants because this will ultimately benefit agriculture. USDA should also, of course, support fundamental research on animals; however, at present a big "push" is required on plants because basic knowledge needed for applications is sorely lacking.

NSF's plant biology and biotechnology-related grants currently average about $70,000 a year for a 2- to 3-year period, a higher level of research support than the average $46,200 a year of USDA's Competitive Grants (see Table 3-2). In addition, NSF supports research on plant biology and biotechnology through several special programs. These programs include postdoctoral fellowships in plant biology, a summer course in plant molecular biology, and the Presidential Young Investigator Awards supporting outstanding young faculty scientists, several of whom are in plant biology. Initiatives in biotechnology include individual investigator awards, cross-disciplinary research and training programs, and proposals for several multidisciplinary biotechnology research centers. NSF's total FY86 appropriation over all its directorates for biotechnology was $86.5 million (Intersociety Working Group, 1986). Within the Directorate for Biological, Behavioral, and Social Sciences alone, grants for biotechnology-related research in 1985 amounted to about $72 million, or 29 percent of the directorate's total research funding.

DEPARTMENT OF HEALTH AND HUMAN SERVICES

National Institutes of Health. NIH sponsors basic and clinical biomedical and behavioral research to improve the health of Americans (FY86 appropriation: $4.9 billion). Much of this research (38 percent) is either directly related to biotechnology or contributes to its broad science base. Several of the institutes support fundamental research on animals, plants, insects, microbes, and diseases that have relevance to agriculture. However, these expenditures represent less than 1 percent of NIH's total budget.

Most of this agriculturally relevant research is supported by the National Institute of General Medical Sciences (NIGMS; FY86 appropriation: $428.6 million). NIGMS's mandate covers biomedical research and research training in the cellular and molecular bases of disease, genetics, pharmacological sciences, physiology, biophysics, and physiological sciences. These studies have spillover effects on animal science that also contribute to agricultural biotechnology. NIGMS funds some research on plant systems that contributes important general information about life processes, such as energy production through photosynthesis and nitrogen fixation. In addition, NIGMS funds research on insects that adds to the general understanding of neurophysiology and development. Thus, NIGMS contributes some support to studies of both plants and insects with spillover benefits for agriculture, but these research areas constitute less than 10 percent and 1 percent, respectively, of its budget.

Of the federal agencies that administer peer-reviewed, competitive grants, NIH funds by far the largest number, and in addition, funds them at a significantly higher level of support. In 1986, some 18,786 new, continuing, and renewed grants were being funded by NIH. The typical NIH grant to an individual university scientist in 1986 amounted to about $164,000 per year for combined direct and indirect costs and was funded for 3–3 1/2 years. This level of funding is adequate to maintain a research program focused on biotechnology or related areas. In contrast, a competitive grant from USDA or NSF is considerably lower and rarely enough to be an investigator's sole source of support for research in these areas (see Table 3-2).

Food and Drug Administration. The FDA is responsible for the regulation of food, drugs, cosmetics, medical devices, and biologics—regardless of how the products are made. FDA regulates biotechnology products under the same rules and procedures used for other products, although they may be subject to different testing requirements. The agency conducts some research, but only to support its regulatory mission. Biotechnologies may be used in such efforts. The Center for Foods (FY86 appropriation: $82.0 million) and the Center for Veterinary Medicine (FY86 appropriation: $23.8 million) occasionally support research that could be applicable to agriculture and biotechnologies.

Department of Energy

Within DOE, the Biological Energy Research Division sponsors research to discover and describe biological mechanisms that could be used as the basis of future energy-related biotechnologies. This research relates explicitly to biotechnology for agriculture across the range of studies on plants and microorganisms that the division funds (FY86 appropriation: $11.8 million). Examples of relevant areas funded include photosynthesis, control of plant growth and development, plant stress physiology, plant cell-wall structure and function, plant–microbe interactions, aspects of microorganisms related to bioprocessing and fermentation, and microbial ecology. In FY86, grants made to individual researchers at universities averaged $72,000 per year.

Department of Defense

The U.S. Army allocated about $50 million in FY86 to R&D involving biotechnology, encompassing mainly vaccine development and disease diagnosis and treatment. This research has wide application to animal health programs. Moreover, some Army researchers and public and private laboratories receiving Army contracts are working directly on important diseases of livestock.

Within DOD, the Office of Naval Research obtains or develops worldwide scientific information and necessary services for conducting specialized and imaginative naval research (FY86 appropriation: $340 million). Of this total, $210 million is distributed through peer-reviewed, competitive contracts. In this context, the Office of Naval Research funds basic research on animals, plants, and bacteria through competitive grants in its biology program (FY86 appropriation: $30 million). Basic research in biotechnology is emphasized in this program, making up about $6 million of the funds awarded. Important research areas include biomolecular engineering, biofouling and biocorrosion, degradation of toxic substances, and synthetic rubbers and fibers.

National Aeronautics and Space Administration

NASA's mandate is essentially unrelated to agriculture or biotechnologies; however, the agency does support two small programs that indirectly give some support to research on plants and animals. The Space Biology Program (FY86 appropriation: $2.5 million) conducts research to identify and describe biological systems that are affected by the gravity-free environment of space as well as research to use space as a tool to probe biological questions that cannot be answered on earth. Basic research funded by this program is about evenly divided between plant and animal systems, with an emphasis on the biological effects of microgravity and the interrelationships among plant growth, light, and other environmental stimuli.

The Controlled Ecological Life-Support System Program (FY86 appropriation: $0.8 million) is a small basic research program that focuses on space containment research, including topics from waste management to food production. The program emphasizes using plants as components of life-support systems in space. Biotechnologies may be used to conduct research, but they do not receive special attention.

AGENCY FOR INTERNATIONAL DEVELOPMENT

The total FY86 appropriation for agricultural research within AID was $30 million. Biotechnology research abroad funded by this program fell into three areas: biological nitrogen fixation ($0.2 million), animal vaccines ($0.87 million), and tissue culture ($0.5 million). Thus, 5 percent of AID's agricultural research budget is now devoted to biotechnology. These data are presented for comparison with other agencies listed previously, which support agricultural biotechnology research within the United States.

State Support of Agricultural Research

State governments contribute significant support to agricultural research. States match, and in recent years have consistently exceeded, the contribution supplied by federal formula funds through CSRS (ESCOP, 1984). Table 3-6, Line 1 documents that this situation is continuing: In 1985, the ratio of state appropriations to CSRS formula funds in SAESs was 3.5:1. In addition, states support land-grant universities and their research facilities, many of which also receive competitive grants from USDA, NSF, and NIH for research related to agriculture.

CSRS is USDA's administrative mechanism to funnel financial support to the SAESs, cooperating forestry schools, land-grant colleges of 1890, and the Tuskegee Institute. There are SAESs in every state, usually associated with a university. They bear the cost of sustaining their own scientific expertise, support personnel, and research facilities and equipment within the academic departments of their universities. In a typical college of agriculture, SAES funding accounts for 60 percent or more of total research and academic faculty salaries and 80 percent or more of the total costs of research and academic activities of the faculty.

Universities operate the backbone of the nation's research programs, and states have traditionally been major supporters of universities. The recent report of the White House Science Council Panel on the Health of U.S. Colleges and Universities (1986) states

Since most basic research can rarely be perceived in terms of specific products and services, and given the long-range nature of such research, private industry does not often support a high level of basic research. If one thing has become clear in recent decades, it is that the fruits of basic research provide benefits for all society, frequently in ways not visible initially to any of the participants. It is for these reasons that the federal government has become, and remains, the primary supporter of basic research in this country.

The important point here is that a strong federal support program is the necessary incentive for research that carries large spillover benefits. For example, the benefits of agricultural research carried on in one state often accrue to the farmers and consumers in other states. Much agricultural research is carried out in the state university system. Without compensating federal funding, states cannot be expected to support lines of basic research whose benefits are more national in scope. Thus, for an optimal national investment in agricultural research, there must be a strong federal commitment to match that of the states.

This committee believes that states should strengthen their already significant role in agricultural research and training. State support for programs in agricultural biotechnology at universities and research stations is important because of the benefits biotechnology can bring to both the state and national economies.

TABLE 3-3   Percentage Expenditure per Area
of Agricultural Research by Private Industry

Major Areas of Research	Percentage of Total Expenditures
Biotechnology	7.2
Human food	14.5
Plant breeding	18.1
Pesticides	33.1
Other[a]	27.1
Total	100.0

[a] Includes farm machinery and equipment, biologics, animal nutrition and feeds, plant nutrients, packaging materials, energy research, agricultural economics, natural fiber processing, and tobacco products and processing.

SOURCE: Adapted from the Agricultural Research Institute, July 1985. A Survey of U.S. Agricultural Research by Private Industry, III. Bethesda, Md. Table IV.

Private Sector

Private industry invests approximately $2.1 billion annually in agricultural research. Of this amount, 95 percent is spent on in-house research. Only 5 percent is spent in support of research conducted outside of industry, sponsored through companies' grants or contracts to universities, foundations, or other public or private organizations (ARI, 1985).

The major areas of agricultural research pursued by private industry, ranked by expenditure from highest to lowest, are pesticides, plant breeding, human food, biotechnology, farm machinery and equipment, biologics, animal nutrition and feeds, plant nutrients, packaging materials, energy research, agricultural economics, natural fiber processing, and tobacco products and processing (Table 3-3). However, in the ARI's survey, a high percentage of companies reported doing either no basic research or no research at all. Thus, much of industry depends on the public sector for necessary developments in basic and applied agricultural research.

Private sector institutions conducting biotechnology research themselves fall into several general categories. There are new entrepreneurial biotechnology companies, generally small and focused on only one or just a few research projects in a narrowly defined area. These are often founded by academic scientists funded by venture capital, R&D limited partnerships, and a sale of equity. Some of these companies survive and grow into more established companies, expanding their R&D efforts as their finances and staff increase. On the other hand, many well-established chemical and drug houses have created their own R&D departments focused on biotechnology projects of interest to them. These efforts range from very directed research to fundamental studies in areas important to future biotechnology applications.

Not-for-profit private sector research institutes also exist. Some of these have been extremely effective in fostering high-quality basic research, which supports progress in biotechnology. Noteworthy examples include Cold Spring Laboratory and the Boyce Thompson Institute in New York, and the Carnegie Institution of Washington, with laboratories located in California and Maryland.

Private sector contributions to biotechnology R&D also occur through the multitudinous links that have grown up around collaborative funding, research, development, and marketing arrangements established among different companies and among companies and other private and public institutions. These contributions are discussed further in Chapter 5.

A Summary of Agricultural Research Funding

It is clear that a variety of federal, state, and private institutions support agriculturally relevant research. Combined, they spend slightly more than $4 billion annually for agricultural research in the United States. Private industry's expenditures represent about half this amount ($2.1 billion; ARI, 1985), combined federal and state support of the traditional agricultural research system accounts for $1.9 billion (USDA, 1986), and the balance of about $100 million represents grants from federal agencies such as NSF, NIH, and DOE for agriculturally related research to universities outside the land-grant system. It must be recognized that such an estimate is conditioned by the difficulty of distinguishing expenditures for agriculturally relevant research from other types of research. Further, there is always the possibility of funds being transferred among public and private institutions and then reported by both. However, we feel that such errors are negligible, and a reasonable estimate of the total public investment—both

TABLE 3-4   Total Government Expenditures for Agricultural Research
Reported by the Current Research Information System[a]

Sponsor	Amount (thousands of dollars)	Percent
USDA, in-house	642,248	33.3
State Agricultural Experiment Stations	1,145,957	59.4
Forestry schools	28,534	1.5
Colleges of 1890/Tuskegee Institute	23,019	1.2
Schools of veterinary medicine	56,410	2.9
Other cooperating institutions	29,722	1.5
Small Business Innovation Research Grants	2,101	0.1
Total	1,927,991	100.0

[a] FY85. Columns may not add due to rounding. See Tables 3-5 and 3-6 for breakdowns of federal and state contributions.

SOURCE: U.S. Department of Agriculture, 1986. Inventory of Agricultural Research Fiscal Year 1985. Washington, D.C.

federal and state—for agricultural research in the United States is about $2 billion annually.

Detailed information on how the traditional agricultural research system's $1.9 billion was spent in FY85 is given in Tables 3-4, 3-5, and 3-6. These tables show breakdowns for state and federal expenditures reported through the Current Research Information System (CRIS) on research conducted by the USDA, SAESs, forestry schools, schools of veterinary medicine, and other agriculturally related institutions. Table 3-5 shows expenditures within USDA-operated research groups. Table 3-6 shows expenditures outside federally operated laboratories and details the federal, state, and private contributions for each group. It should be noted that the values are gross figures representing total expenditures, which include costs for administration, rent, and operation of research farms as well as personnel, materials, and other costs related to research.

Of the total $1.9 billion government expenditure reported by CRIS, about one-third is spent on research within the USDA, primarily by the ARS, the Forest Service, and the ERS (Tables 3-4 and 3-5). The SAESs account for about 60 percent of the total expenditures, with state appropriations accounting for more than half of that research support. Federal formula funding through the Hatch Act and other special funding through the CSRS is only about 16 percent of the total research expenditures at SAESs (Table 3-6). Funds from USDA and other federal agencies provided through grants, contracts, and cooperative agreements account for just over 10 percent of the support to SAESs, and approximately equal funding is provided to SAESs by industry and other private sources. It is interesting to note that income from the sale of agricultural products, such as dairy products and meat, is reinvested to support research.

The CRIS inventory also lists scientist-years, a measure of the time scientists devoted to this research. Work by laboratory technicians and graduate assistants and time spent in research administration are not included in tabulating scientist-years. In 1985, the $1.9 billion of state and federal expenditures represented the work of 11,133 scientists, or an average cost of $173,000 per year to support a research scientist. However, this figure is just an average. Some areas of research require less support for equipment, facilities, chemicals, and other expenditures, whereas other areas require much more. Equipment and materials to carry on research in biotechnology are generally more expensive than those for other areas of agricultural research. Hence, biotechnology research probably requires more than the average $173,000 per scientist per year. In private industry the calculation of the per-scientist cost for agricultural research is $159,756 (ARI, 1985). However, this average includes all scientists, whether they hold a B.S., M.S., or Ph.D., in contrast to the CRIS definition of a scientist, which

TABLE 3-5   USDA In-House Expenditures for Agricultural Research Reported by the Current Research Information System[a]

Sponsor	Amount (thousands of dollars)
Agricultural Cooperative Service	2,071
Agricultural Research Service	470,442
Economic Research Service	46,405
Forest Service	118,240
Human Nutrition Information Service	5,090
Total	642,248

[a] FY85. Columns may not add due to rounding.

SOURCE: U.S. Department of Agriculture, 1986. Inventory of Agricultural Research Fiscal Year 1985. Washington, D.C.

TABLE 3-6 Government Expenditures (thousands of dollars) Outside USDA for Agricultural Research Reported by the Current Research Information System[a]

Sponsor	CSRS[b]	Competitive Grants	Other USDA	Other Federal	State Appropriations	Product Sales	Industry and Other	Total
State Agricultural Experiment Stations	181,978	7,729	31,038	90,318	644,461	65,322	125,111	1,145,957
Forestry schools	2,981	112	3,473	4,376	12,549	321	4,722	28,534
Colleges of 1890/Tuskegee Institute	22,310	0	270	146	65	216	12	23,019
Schools of veterinary medicine	2,224	3	905	17,025	21,212	4,636	10,405	56,410
Other cooperating institutions	41	23,178	5,643	549	5	NA	306	29,722
Small Business Innovation Research Grants	NA	2,101	NA	NA	NA	NA	NA	2,101
Total	209,534	33,123	41,329	112,414	678,292	70,495	140,556	1,285,743

NOTE: Columns may not add due to rounding. NA = not applicable.

[a] FY85.

[b] Cooperative State Research Service. Includes funds under Hatch Act, McIntire-Stennis, Evans–Allen, Animal Health, Special Grants, and other CSRS programs.

SOURCE: U.S. Department of Agriculture, 1986. Inventory of Agricultural Research Fiscal Year 1985. Washington, D.C.

includes only the Ph.D. or equivalent level. Another survey of private industry by the Committee on Biotechnology of the Division of Agriculture of the National Association of State Universities and Land-Grant Colleges found a range of from $80,000 to $500,000 per principal (Ph.D.) scientist in agricultural biotechnology R&D, with an overall average of $160,967 (NASULGC, 1985).

PEER REVIEW

Throughout this report, we stress the importance of a peer-reviewed, competitive process for allocating most research funds. Peer review is one of the most effective mechanisms available to ensure that public dollars are invested in relevant, high-quality research and that judgments made in allocating funds are equitable and discerning (Cole et al., 1978).

Peer review, which in its broadest form is also called merit review, can take a variety of forms and serve a variety of purposes. Review by experts is critical to evaluative decisions such as judging the relevance and quality of proposed research, judging the merit of papers to be published, measuring the quality of people for decisions on promotion within universities and research facilities, and setting the direction and priority of research. Review involving other qualified researchers also provides a forum for scientific communication and advice. All research activities should undergo peer and merit appraisal of their scientific worth. In addition, whenever open competition is appropriate to meet the objectives of a program, this evaluative process should be used to distinguish among competitors.

Participants and procedures in the review process should be organized to match the nature of the tasks. A system of review and awards should work to ensure equal opportunity among investigators, a minimum of errors, fairness, and that the best research is selected that can at the same time be managed with reasonable costs in time and money. No system of review can be totally free of error, differences of judgment, or personal preferences (Cole et al., 1978). However, careful attention to the quality and breadth of expertise represented on review panels is the best way to ensure the soundness of their recommendations. Panels should not be composed entirely of people who have a substantial interest in the outcome of the research. These panels should include experts from outside the specific discipline as well as people from another level of the research process, in order to evaluate both the merit and scientific quality of proposed programs. For example, reviewers of a proposed basic research study should include representatives of applied research, and similarly, applied research proposals should be reviewed by some individuals with strong basic research backgrounds. Efforts must be made to broaden the expertise represented on review panels, so the panel can fully evaluate the quality and relevance of proposed research and minimize bias. In addition, objective selection and frequent rotation of reviewers is desirable, to avoid creating an "old boy" network and to bring fresh insights to review panels.

Recent budget constraints are not a short-term phenomenon; the scientific community must operate on the assumption that there will be no real growth in basic research budgets until perhaps the end of the century (Press, 1986). Thus, science has an increasingly important obligation to ensure the optimal use of limited funds. Evaluation and competition through peer and merit review is the most appropriate mechanism to accomplish this goal effectively and fairly. This form of quality control has proven itself through many years of service in the biomedical and basic science research communities, as evidenced by the success of NIH- and NSF-supported research programs. A peer and merit review process must be used to assess and guide the development of the agricultural biotechnology research system. Implementation of these review processes will vary depending on the activity under review, for example, competitive research grants, appropriated formula funds, or agricultural extension. In all cases, the benefits of peer and merit review—properly done and heeded—are continuous monitoring of research advances, more efficient, relevant, and higher quality research, and increased communication and respect among scientists.

REALIGNING THE SYSTEM FOR BIOTECHNOLOGY

Biotechnology requires a large initial investment in what is traditionally referred to as basic research. An understanding of the physiology, biochemistry, and genetics of a biological process is needed before one can use the tools of biotechnology to control that process. Basic research questions are often a necessary component of resolving agricultural problems using biotechnology. Thus, as in many other areas of science, there is substantial overlap between basic and applied research. Despite past institutional arrangements and funding patterns that emphasized the separation of basic and applied research, biotechnology is bringing agricultural research closer to Pasteur's dictum: "There is no pure science and applied science, only science and its applications." The agricultural research system in the United States must better integrate basic and applied research as it moves to facilitate the advances biotechnology can make for agriculture.

Funding for Agricultural Biotechnology

Current expenditures for biotechnology research in the agricultural research system cannot be documented or compared with any precision, because few analyses of biotechnology research support were done until very recently. In addition, there is no widely accepted definition of biotechnology, which makes it difficult to establish clear-cut criteria for classifying such research. However, some general estimates are available from both the U.S. General Accounting Office (GAO, 1985) and the National Association of State Universities and Land-Grant Colleges (NASULGC, 1983). Both of these reports show that biotechnology research represents only a small part of agricultural research funded through USDA.

The NASULGC study reported data collected through a questionnaire mailed to SAESs in 1982. Total support for biotechnology research at SAESs was $41.5 million, and the survey reported that this funding came from state ($16.2 million), federal ($19.8 million), and private ($5.5 million) sources. The GAO study listed data collected for FY84, from another survey of SAESs and Colleges of Veterinary Medicine, which put total support for biotechnology research at $47.2 million (see Table 3-7 for a breakdown on sources of funding). In addition, the GAO (1986) reported that support for biotechnology research within ARS in FY85 was $24.5 million, or 5.2 percent of its total research budget.

TABLE 3-7   State Agricultural Experiment Station and Veterinary
College-Sponsored Biotechnology Research (millions of dollars) for FY84[a]

Source of Funding	Biotechnology Research	Total Agricultural Research	Percentage of Total Agricultural Research that is Biotechnology Research
Competitive Research Grants Office	2.8	11.6	23.9
All other USDA funds	7.9	173.6	4.6
Other federal agencies	13.6	109.1	12.5
State agencies	17.3	551.2	3.1
Industry	5.6	86.3	6.5
Total	47.2[b]	931.8	5.1

[a] Relates to the 495 research projects discussed in the Government Accounting Office (GAO) Report; data for FY84.

[b] This figure does not include an estimated $500,000 reported to the GAO by the North Dakota Agricultural Experiment Station or $1,693 reported by the Ohio Experiment Station. These two stations, although providing GAO with a total figure for their biotechnology research, did not identify the specific sources of that funding and GAO, therefore, excluded the amounts from the table.

SOURCE: General Accounting Office, 1985. Biotechnology: The U.S. Department of Agriculture's Biotechnology Research Efforts. Washington, D.C. (GAO/RCED-86-39-BR).

These data make an interesting point: Biotechnology research at the ARS, SAESs, and veterinary colleges accounts for approximately 5 percent of total research funding at these institutions. However, a larger percentage of the grant support provided to SAESs and veterinary colleges by the USDA competitive grants program and federal agencies other than the USDA goes for biotechnology research. In fact, 12.5 percent of support from other federal agencies and 23.9 percent of USDA competitive grant support to SAESs funded biotechnology research (Table 3-7).

As the awareness of biotechnology's role in research increases, government agencies have begun to track their expenditures in biotechnology (Table 3-8). Categorizing what biotechnology research actually is can be somewhat arbitrary, because biotechnology methods are being used in almost all biological disciplines and in some areas of engineering and chemistry. In addition, federal agencies have not formally agreed upon a definition for biotechnology; both narrow and broad criteria are used, which limits the significance of comparing levels of funding among agencies. And though direct comparison of the dollar values is not valid, funding

TABLE 3-8 A Comparison of Data on Funding Levels Available for FY84 and FY85 on Biotechnology and Agriculturally Related Biotechnology Research by Selected Sources

Sponsor	Amount (millions of dollars)
AGRICULTURALLY RELATED BIOTECHNOLOGY	
USDA[a]	
Agricultural Research Service	24.5
Cooperative State Research Service:	
Competitive Research Grants Office	30.0
Hatch Act and Special Grants	18.4
SAES (nonfederal support)[b]	
State funding	17.3
Industry	5.6
Private industry[c]	150.0
ALL BIOTECHNOLOGY[a]	
EPA	1.5
FDA	2.6
NIH	1,849.5
NSF	81.6

NOTE: EPA = Environmental Protection Agency; FDA = Food and Drug Administration; and SAES = State Agricultural Experiment Stations.

[a] FY85. Competitive Research Grants Office funding includes both specific biotechnology grants and additional biotechnology-related research covered by its other grants. Funding by non-USDA federal agencies may include some agriculturally related biotechnology research. SOURCE: Government Accounting Office, 1986.

[b] FY84 data.

[c] Estimate based on data from the Agricultural Research Institute, 1985. A Survey of U.S. Agricultural Research by Private Industry III. Bethesda, Md.

SOURCE: National Association of State Universities and Land-Grant Colleges, 1985.

information does give some indication of what different government agencies estimate they spend on biotechnology (see Tables 3-2 and 3-8).

Ultimately, the important consideration is the availability of adequate funding to support significant advances in biotechnology. What does it cost to make progress in agriculturally related biotechnology? The following is one estimate of the price tag on a discovery in biotechnology of value to agriculture.

DEVELOPING A DISCOVERY INTO A RESEARCH TOOL:
THE COST OF THE AGROBACTERIUM TI PLASMID

How much does it cost to take a discovery in molecular biology and develop it into a useful biotechnology? To arrive at an answer, other questions must be considered. For instance, how many scientists are working, in how many laboratories, and over how many years? How do you account for the related basic knowledge that laid the foundation for the discovery? How do you define what other variables are involved in calculating the true costs?

The *Agrobacterium* Ti plasmid is one of the earliest biotechnology success stories in plant research and is a classic example of how happenstance combines with years of effort to provide a useful research tool. The route to the discovery began at the turn of the century, with research on a plant disease called crown gall. USDA scientists discovered that *Agrobacterium tumefaciens* was the disease agent. By the 1940s, about 20 scientists concentrated in three laboratories (one in the United States and two in France) were actively studying fundamental aspects of the disease. By the late 1960s the worldwide effort had grown to include about 40 researchers in 10 different laboratories.

At first, the work was of interest to only a small group of people studying plant diseases. Then in 1979, following the discovery that the bacterium was actually transferring genetic material to higher plants, the research effort exploded. Scientists quickly saw the practical potential of this mechanism for gene transfer. About 40 scientists worked in 10 laboratories for 4 years reconstructing the Ti plasmid as a plant gene transfer system. Throughout the early 1980s, laboratory studies related to plant gene transfer and to the Ti system occupied the talents of up to 250 additional scientists. By 1986, at least 300 people working in about 25 laboratories worldwide were conducting research on both applied and fundamental aspects of the Ti plasmid system. The annual estimated cost of this research worldwide was about $45 million. (This amount assumes an average expense of $150,000 per scientist per year.)

Adding up the costs of the research directly related to the development of the Ti plasmid gene transfer system gives only a general estimate of the expense of developing one technical breakthrough in biotechnology. Much of the research using the Ti plasmid in plant gene transfer is being supported either by private industry or by competitive grants to universities.

THE FEDERAL ROLE

As stated earlier, the commitment to basic research is key to applying the promise of biotechnology to agriculture. Future directions and applications, as well as new technologies, will emerge from fundamental studies of metabolic pathways and the regulation of growth and development funded by federal research agencies. Private industry and state governments cannot be expected to invest significantly in such long-term, high-risk research. Upfront investment in the future of agricultural biotechnology is a federal responsibility.

Clearly, USDA has an obligation to step up its support of biotechnology research. USDA could increase its emphasis on biotechnology in two ways: by adding more money or by redirecting existing money. Any increase in funding at USDA should not come at the expense of appropriations to other federal agencies that support research relevant to agriculture. Redirection of some existing research program funds must also occur within the USDA budget to heighten the priority given to biotechnology. This redirection can be done most effectively by a substantial increase in research awards through the Competitive Research Grants Office Program.

Greater emphasis is needed on agricultural biotechnology within both the USDA and the NSF to maintain the nation's competitive position in agriculture, technology, and world markets. Given the current average cost of $173,000 per year to support a research scientist at an SAES and a projected demand for 3,000 active scientists working in biotechnology research related to agriculture (see Chapter 4), federal funding should be increased in this area to about $500 million per year by 1990. This support should be administered by the primary federal agencies supporting agricultural biotechnology (USDA and NSF) in the form of peer-reviewed, competitive grants.

Integration of Agricultural Research Disciplines

Agricultural research depends on basic science, applied science, technology development, and technology transfer (including extension). In realigning the research system to promote advances in biotechnology, communication must be maintained among basic researchers, applied researchers, and the farmers and private companies who use the technology. If the system is to be effective, we must strengthen both the links among disciplines of science supporting agriculture and the links between basic and applied research and technology development and transfer.

Integration of different disciplines is important because it facilitates the blending of skills and knowledge. For example, the fields of biology and chemistry have been integrated in biochemistry. Cytology and genetics have come together to provide new insights into gene identification. In addition, the already hybrid fields of biochemistry and chemical engineering have joined forces in developing bioprocess and fermentation technology. Integration of basic and applied research and technology development and transfer is particularly important in biotechnology because this field has developed from the confluence of basic science and technology development.

Integration of research from basic science, to applied science, to technology development, and then to technology transfer has traditionally been carried out by land-grant universities, and these institutions will continue to play an important role in the future. Yet new institutional forms are now emerging outside the traditional land-grant system as efforts mount to improve efficiency in the development of profitable technology. These new forms of integration are being encouraged in part by the rapid growth of private sector research in biotechnology.

LAND-GRANT UNIVERSITIES

Land-grant universities are well suited to foster the integration of research to develop and apply biotechnology because of their tripartite structure—teaching, research, and extension. Land-grant universities with strong basic science departments are able to mount a continuum of activities ranging from fundamental research, to applied research, and then to extension. Cooperative extension provides a feedback mechanism to let researchers know whether the technologies they develop are appropriate to the needs of their clientele. Because of federal budget cuts in formula funding for both research and extension, the research programs of these land-grant universities depend increasingly on financing through competitive grants from both public and private sources. Although this increased dependence on grants should improve the quality of scientific research, feedback between the clientele and scientists has been weakened considerably. Thus, both the quality of research and its relevance to the end users must be taken into consideration in the research review process.

To foster the integration of research, there must be an environment within the university that encourages cooperation across departments and colleges, and across basic and applied research entities. A key to this environment is the recruitment of high-quality faculty in all areas. The reward system of the university should also be responsive to and supportive of integrated programs if these are to succeed.

Integration in agricultural research should be promoted and supported. Universities need to establish graduate programs that cut across departmental lines; recognize and reward faculty contributions to cooperative research programs; promote collaborative projects and exchanges between researchers in land-grant universities, non-land-grant schools, industry, and government laboratories; and recruit faculty to create interdisciplinary research teams that can attract competitive funding.

NEW INSTITUTIONAL FORMS

New institutional forms can be created to help facilitate the integration of biotechnology research. One example is the creation of centers focused on one or more specific agricultural issues. The publicly supported Michigan Biotechnology Institute (discussed in Chapter 5) is one example of a center that integrates basic and applied research. This center, located near Michigan State University, focuses on the applications of biotechnology to renewable resources that benefit the state. It conducts both basic and applied research aimed at developing and patenting new technologies and products. If this organization is successful, similar institutions are certain to develop.

Other linkages are being established between applied research institutions or businesses and basic research centers. For example, the Rockefeller Foundation is funding a program on biotechnology for rice, which will link the work of scientists at the International Rice Research Institute in the Philippines with that of scientists in basic research laboratories in the United States and Europe. The seed company Pioneer Hi-Bred International, Inc. has given a grant to Cold Spring Harbor Laboratory in New York and will station one of its scientists at this basic research institute. Other collaborative research programs, such as the Cornell University Biotechnology Program (discussed in Chapter 5), link private company researchers and university basic research programs.

New Approaches to Agricultural Biotechnology

Several steps could be taken to encourage the integration of research. Federal and state governments should support the establishment of collaborative research centers, promote interdisciplinary conferences and seminars, support sabbaticals for government scientists and other exchange and retraining programs with universities and industrial laboratories, and provide funding for interdisciplinary project grants.

Grants for Interdisciplinary Research

In biomedical sciences and human health, it is not uncommon for articles published in scientific journals to have a half dozen or more coauthors. Multiple authorship often reflects productive interdisciplinary collaboration. In the agricultural sciences, the tradition of individual achievement is still strong. There should be a significant increase in grants designed to encourage interdisciplinary research, such as those sponsored by the McKnight Foundation (see Chapter 4).

Collaborative Groups and Exchanges

The land-grant universities potentially have a strong capacity for interdisciplinary and collaborative research efforts in agricultural biotechnology. Private universities, in contrast, have few agricultural science-related disciplines. However, private universities do have reservoirs of talent in basic sciences that are essential for biotechnology development. It would be highly advantageous for the development of agricultural biotechnology to promote both long-term collaborations and temporary exchanges among land-grant and other public and private research universities. For instance, USDA agronomists and other agricultural scientists should be encouraged to take sabbaticals at non-land-grant institutions. Longer term collaborative projects between land-grant and non-land-grant institutions would help pave the pathways of information exchange.

Exchange of personnel between public-sector research institutions and private companies engaged in research should also be encouraged. Research in the private sector tends to have a stronger focus on teams, and the reward system is often more conducive to interdisciplinary research.

LARGE LABORATORY GROUPS

Large, autonomous laboratory groups can also function effectively to pursue some biotechnology-oriented research goals. Such groups are especially needed at universities that have limited faculty in areas such as plant science. A large laboratory with 15 or more scientists will have the manpower and resources to attack research problems that cannot be effectively handled by small laboratories or by individual scientists working in isolation.

RESEARCH CENTERS

The NSF has been instrumental in setting up 11 Engineering Research Centers, each of which is based at a university selected through rigorous competition. These centers receive substantial funding from industry as well as from the federal government. They bring together academic and industrial researchers to attack specific scientific problems in a multidisciplinary setting. Examples of this approach initiated in biotechnology include MIT's Biotechnology Process Engineering Center and Cornell's Biotechnology Research Program. The center concept can be extended to integrate basic science and technology development activities.

RECOMMENDATIONS

LINKING AND INTEGRATING RESEARCH

The tools and approaches of biotechnology are equally relevant to science-oriented research and technology-oriented research. Biotechnology can strengthen as well as benefit from improved linkages between basic scientific research and research to adapt technology to agricultural problems. Equally important, different disciplines within biology and agriculture can collaborate to integrate knowledge and skills toward new advances in agriculture.

New approaches to agricultural research are needed to establish strong and productive linkages between basic science and its applications as well as interdisciplinary systems approaches that focus a number of skills on a common mission. Just as biochemistry, genetics, molecular biology, and fields of medicine have successfully joined forces to solve medical problems, integration of these scientific disciplines for agricultural research must be promoted and supported by appropriate recognition and reward through university, industry, and government channels.

First, universities should establish graduate programs that cut across departmental lines; recognize and reward faculty contributions to cooperative research programs; promote collaborative projects and exchanges between researchers in land-grant universities, non-land-grant universities, industry, and government laboratories; and recruit faculty to create interdisciplinary research programs that can attract competitive funding. Faculty should be selected by departments or groups representing two or more disciplines (e.g., genetics and entomology or biochemistry and botany).

Second, federal and state governments should support the establishment of collaborative research centers, promote interdisciplinary conferences and seminars, support sabbaticals for government scientists and other exchange and retraining programs with universities and industrial laboratories, and provide funding for interdisciplinary-program project grants.

PEER AND MERIT REVIEW

A peer and merit review process must be used to assess and guide the development of the agricultural biotechnology research system, including all steps from basic science to extension.

The participants and procedures in the review process should be organized to match the nature of the tasks and programs reviewed and must include individuals outside the organization as well as experts from relevant disciplines and from basic and applied research programs.

Efforts must be made to broaden the expertise represented on review panels in order to best examine the quality and relevance of work with minimal bias. The benefits of peer and merit review—properly done and heeded—are continuous monitoring of research advances; more efficient, relevant, and higher quality research; and increased communication and respect among scientists.

THE FEDERAL GOVERNMENT'S ROLE

It is logical that primary funding for agricultural biotechnology should be achieved through the USDA. Unfortunately, funding for both intramural and extramural basic research within USDA is well below that of other federal agencies. USDA has recognized the need to support basic research and is attempting to do so, albeit not as rapidly as might be optimal. Funding increases are needed. Allocation of new and even redirected funding should be based principally on competitive peer and merit review.

Any increase in funding at USDA should not come at the expense of appropriations to other federal agencies that support biological research relevant to agriculture. This is because it is not always clear where innovation applicable to agricultural biotechnology might arise. However, some existing research program funds should be redirected within USDA to heighten the priority given to biotechnology. USDA should also emphasize related fundamental research on animals and plants, the lack of which is impeding the application of biotechnology to livestock and crop improvement.

Funding for competitive grants through USDA must be of a size and duration sufficient to ensure high-quality, efficient research programs. The recommended average grant should be increased to $150,000 per year for an average of 3 years or more. This level of funding is consistent with the current average support per principal investigator used by industry and USDA/ARS intramural research programs. The duration of these competitive grants is also in accord with the recent recommendation:

Of equal importance with the level of funding is the stabilization of federal support to permit more effective use of financial and human resources. . . . Federal agencies [should] work toward an average grant or contract duration of at least three, and preferably five, years. (White House Science Council, 1986)

The committee recommends that competitive grants by all agencies in the federal government for biotechnology research related to agriculture total upwards of $500 million annually, a level that could support 3,000 active scientists. This level of support should be achieved by 1990, primarily through competitive grants administered by USDA and NSF.

THE STATE GOVERNMENTS' ROLE

States should continue to strengthen their already major role in agricultural research and training through their support of universities and research stations that conduct regional research. They should continue to focus on identifying regional interests and on supporting the training of personnel needed in agriculture. The states should also evaluate programs in agricultural biotechnology and the role such programs can and will play in each state's economy.

THE PRIVATE SECTOR'S ROLE

The private sector's traditional emphasis on product development is not likely to change, even though there has been a dramatic increase since 1980 in private sector investment in high-risk basic research in agricultural biotechnology. Because public sector investment provides skilled manpower and the knowledge base for innovation, industry should act as an advocate for publicly supported training and research programs in agricultural biotechnology. Industry can also support biotechnology research through direct grants and contracts to universities, cooperative agreements with federal laboratories, and education to inform the general public about the impacts of agricultural biotechnology.

Foundations should be encouraged to support innovative science programs in order to maximize their potential for having substantial influence in important areas. The McKnight Foundation's interdisciplinary program for plant research and the Rockefeller Foundation's efforts to accelerate biotechnology developments in rice are noteworthy examples. Other foundations should address equally important experiments in technology transfer and extension for agricultural biotechnology.

4
Training

INTRODUCTION

To initiate and implement advances in biotechnology for agriculture will require more than appropriate institutional structures and funds. A strategy for biotechnology also requires a work force of agricultural research scientists trained to apply molecular biology techniques critical to solving agricultural problems. Because biotechnology research spans a continuum from basic science to practical application, its practitioners must be conversant with the general biology of an organism and with the biochemical and genetic details of its life cycle. This new breed of researcher must understand the techniques of molecular biology and possess the skills to modify these techniques to suit particular organisms.

Scientists, administrators, faculty, and policymakers should be aware of the importance of sound education and training to the progress of agricultural biotechnology. Programs are needed to attract young scientists to modern agricultural research and to effectively train these scientists. Other programs are needed to retrain traditional agricultural scientists in biotechnological methods, so they can apply these powerful tools to two of the long-standing goals of agricultural research: improving food quality and production efficiency. Four types of programs merit increased federal support: pre- and postdoctoral fellowships, training grants, career development awards, and retraining opportunities. To be most effective, these programs should be administered on a peer-reviewed, competitive basis.

Training in agricultural sciences shares many features with training in other biological sciences. Therefore, for this report the committee examined existing biological and biomedical training programs that could prove helpful in improving programs for agricultural sciences. The following section discusses issues relevant to training in agricultural biotechnology.

PERSONNEL REQUIRED FOR BIOTECHNOLOGY

Demand for Scientists

As biotechnology industries grow, so grows the demand for trained employees. Scientists competent in biotechnology tools are in increasing demand not only by the industries serving medicine and agriculture but also by the educational system. Academic institutions need to expand the expertise of their faculty to provide training in biotechnology. Several recent independent estimates substantiate this demand.

Although it is difficult to know the precise number of trained personnel required by the U.S. biotechnology industry, which listed 1,471 companies in business in 1985 (Sitting and Noyes, 1985), employment in the biotechnology industry has been increasing by about 25 percent per year since 1973 for Ph.D.s in biomedicine (Institute of Medicine [IOM], 1985). Based on a survey, the IOM estimated that 12,000 scientists were employed by the U.S. biotechnology industry in 1985 and that just under half of them (4,000–5,000) had Ph.D.s (IOM, 1985).

There are indications that the number of scientists practicing biotechnology in agriculture has also increased. A 1983 survey by the Office of Technology Assessment (OTA) of 219 selected biotechnology companies showed that 28 percent of these companies were working in animal agriculture and 24 percent in plant agriculture (OTA, 1984). Clearly, agriculture is a major beneficiary of biotechnology research.

The unmet demand for researchers in agricultural biotechnology is substantial. In a survey conducted by the American Council on Education (ACE), academic, government, and industry laboratories all cited shortages of personnel trained in plant molecular biology, biochemistry, and genetics (Anderson, 1984). In addition, the 1985 IOM biotechnology questionnaire found that firms looking for plant scientists cited shortages of (1) plant molecular biologists with solid training in plant science (vs. training in bacterial or animal systems); (2) plant tissue culture experts; (3) plant geneticists or breeders with expertise in a second area such as tissue culture, cell biology, or molecular biology; and (4) bioprocess engineers (IOM, 1985).

A survey by the Division of Agriculture Committee on Biotechnology of the National Association of State Universities and Land-Grant Colleges (NASULGC) projected 35 percent growth for Ph.D.-level scientists and 53 percent growth for B.S.-/M.S.-level scientists in agricultural biotechnology over the next 3 years (NASULGC, 1985). Companies cited many areas of expertise they required, but comments were also made about the importance of a broad education to cope with rapidly changing science and the need for agriculturalists who understand techniques in biotechnology. The survey reported that 1,244 scientists were working at 38 responding companies. Assuming 35 percent growth, 1,678 scientists would be needed by these companies alone by 1987. Our committee has estimated that at least 3,000 scientists are now needed in the public agricultural biotechnology sector.

The IOM has pointed out that "while industry may provide an increasing share of employment opportunities . . . universities will still be counted on to provide most of the training" (IOM, 1983). This statement applies to agricultural biotechnology, but as noted in a recent report by the National Agricultural Research and Extension Users Advisory Board (1986):

Basic science curriculums in colleges of agriculture must be brought up to the same standards as those in the colleges of science. Many agricultural colleges offer courses to agriculture majors in the basic sciences that are not as stringent as those offered by colleges of science.

Although the demand for trained personnel is clearly growing, only a few federal training programs exist to help fulfill these needs. To make progress in agricultural biotechnology, federal support for graduate education must be increased to ensure the future supply of scientists. In addition, increased postdoctoral opportunities in agricultural research are needed to attract, train, and keep young Ph.D. scientists in agricultural biotechnology.

The Report of the White House Science Council Panel on the Health of U.S. Colleges and Universities (1986) noted, "the federal government is the primary supporter of basic research in this country." The report called for "a substantial program of merit-based, portable scholarships . . . by the federal government at the undergraduate level. . . . [and] Substantial programs of multiyear merit-based fellowships . . . at the graduate level." These types of scholarships and fellowships are needed in many fields. Allocating some of them to modern agricultural research would help to ensure the nation's supply of scientists trained in agricultural biotechnology.

Demographic Trends

Another justification for increasing federal support to graduate education in agricultural biotechnology—as well as science in general—is demographic: The college-age population is declining, and this decline will decrease the pool of graduate students and could lead to a shortage of research personnel (Figure 4-1). In analyzing the effects of demographic factors on biomedical research, IOM recommended maintaining federal support of graduate training to offset possible future shortages of research personnel (IOM, 1985).

Graduate education not only produces scientists, it also contributes to U.S. research productivity by the experimental work students perform for their Ph.D. theses and by their later research as mature scientists. Graduate students cannot be replaced by technicians, who usually do not design experiments or train to become research leaders. If the number of graduate students declines because of demographics (e.g., a declining college-age population pool) and decreased funding (for both educational and research programs), research productivity will suffer unless there is a compensating rise in the number of postdoctoral researchers. However, because postdoctoral researchers are supplied from the graduate student pool, their numbers will likely shrink as well.

A decline in the college-age population will have another effect on training: Decreased revenues from tuition will support fewer permanent faculty. IOM has concluded that biomedical science faces a long-term prospect of fewer graduate students, more postdoctoral researchers with longer term, semipermanent positions,

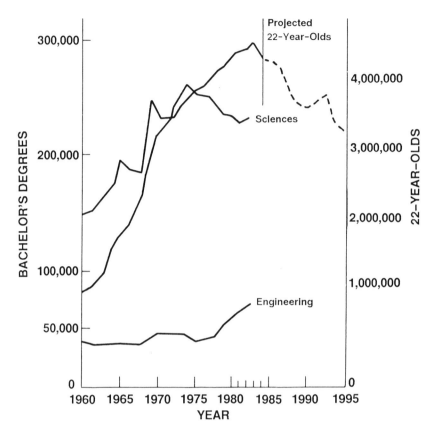

FIGURE 4-1   Science and engineering Bachelor's degrees and the 22-year-old population. Source: Bloch, E. 1986. Basic Research: The Key to Economic Competitiveness (Fig. 10, p. 9). Washington, D.C.: U.S. Government Printing Office.

and more technicians (IOM, 1983). Because these conclusions are likely to apply to agricultural science as well, it is important to note the status of postdoctoral study in the two fields. In 1983, 59 percent of new Ph.D.s in the biological and health sciences planned postdoctoral study, contrasted with 18 percent of agricultural science Ph.D.s (National Research Council [NRC], 1983). These figures reflect the fact that postdoctoral training in biomedicine is considered a necessary transition between graduate education and a faculty or equivalent position, but the same is not generally true for the agricultural sciences (Anderson, 1984). In bypassing postdoctoral study, agricultural scientists may receive tenure faster, but at the same time they may find that they have limited exposure to modern developments in research.

Another notable demographic trend involves the number of foreign students in the United States. The 1985 Science Indicators report (National Science Board [NSB], 1985) shows an increase in the percentage of foreigners receiving doctoral degrees in the United States for most scientific fields (Figure 4-2). The trend is attributed to both a decrease in the number of U.S. students earning Ph.D.s and an increase in the number of foreign graduate students in the United States. The high percentage of foreigners earning Ph.D.s in agriculture in the United States is noteworthy. In 1983, 20 percent of all graduate students and 33 percent of all postdoctoral researchers in plant biology in the United States were foreigners (Anderson, 1984). Half of the foreign graduate students and one-third of the postdoctoral researchers received major support from their governments. These researchers are highly productive while training in the United States. In addition, many continue their scientific careers in this country, rather than returning home on completing their training.

More training opportunities and incentives are needed to attract U.S. students and scientists to work on problems in agriculture. Without these new opportunities and incentives, the United States runs the risk of losing its leadership role. Therefore, federal support for training U.S. scientists in agricultural biotechnology must be increased.

EDUCATION AND TRAINING

Sound, comprehensive education is a prerequisite for scientific training. Furthermore, breakthroughs are often made by scientists who may specialize in unrelated fields but have a breadth of knowledge and an appreciation of several disciplines. Such individuals can bring fresh insights to bear on research problems. Some of these scientists will also act as innovators and guide important transitions in research.

Numerous studies have documented problems in American scientific education: High school and college students show declining test scores in science and mathematics, the academic competency

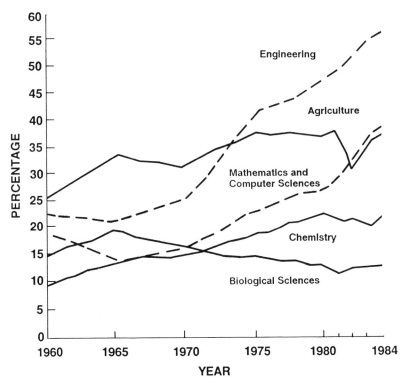

FIGURE 4-2 Doctoral degrees awarded to foreign students as a percentage of all doctoral degrees granted by U.S. universities, by field. Source: Adapted from National Science Foundation. 1985. Science Indicators. Washington, D.C.: U.S. Government Printing Office.

of many science and mathematics teachers is questionable, and patterns of undergraduate majors are changing—that is, 50 percent fewer arts and sciences degrees are awarded compared with business and management degrees (National Center for Education Statistics, 1985). The NSB has recommended several ways to upgrade the quality of science education in America, including increasing science and mathematics instruction in secondary schools and raising college entrance requirements in science and mathematics (NSB Commission on Precollege Education in Mathematics, Science and Technology, 1983). We agree that these actions are needed, as well as an earlier and greater emphasis on science and agriculture in elementary, junior high, and high schools.

Agricultural research must be able to attract top-quality scientists. Appropriate institutional structures and funding patterns can help make agricultural research a more attractive career. However, a sound scientific education should begin at the undergraduate level, when students are taught the fundamentals of the many basic disciplines that underlie biotechnology. These disciplines include chemistry, biochemistry, genetics, physiology, and cell and developmental biology. Students need rigorous education in these basic sciences if they hope to go on to graduate study and later research using the sophisticated techniques of biotechnology.

Specialized training in narrow research areas is more appropriate to graduate and postdoctoral work, after students have acquired the breadth of knowledge that allows them to think creatively about research problems. Thus, there is a distinction between education and training. Both are essential to the progress of research, but the latter cannot be effective without the former.

Several types of programs can attract and train top-quality scientists for careers in agricultural biotechnology research. Programs can address needs at several stages of research training: education at pre- and postdoctoral levels, developing careers in research for young faculty, retraining established agricultural scientists to use biotechnological techniques, and facilitating interdisciplinary projects that are critical to the success of biotechnology. Industry could play a more active role in retraining scientists by initiating and funding courses and collaborative projects.

Federal and state funding of university laboratories is instrumental in training scientists. Clearly, more training in biotechnology must be provided by agricultural schools to fulfill the personnel requirements of academic, government, and industry laboratories. However, the ability to attract graduate students and faculty to agricultural molecular biology and biotechnology will depend on a perception that job opportunities exist and that funding is available. Current U.S. Department of Agriculture (USDA) competitive grants average $50,000 per year for 2 years, and National Science Foundation (NSF) grants in plant sciences average $70,000 per year for 2-3 years. However, applicants to both programs have a success rate of only 15-20 percent. This fact, and the low level of funding compared to National Institutes of Health (NIH) grants, does not encourage students or faculty to enter the field of agricultural biotechnology. In the mid- and long term, this situation could hurt the United States' competitive advantage in biotechnology and its application to agriculture. The following section discusses the status of programs in government agencies that include training in agricultural biotechnology.

Programs at the U.S. Department of Agriculture

PREDOCTORAL

In 1984, the USDA initiated a program with $5 million that supported 302 predoctoral students through peer-reviewed, competitive training grants awarded to university departments. These training grants covered four areas (each area's share of funds is given in parentheses): food science and human nutrition (20 percent), agricultural engineering (20 percent), food and agricultural marketing (25 percent), and biotechnology (35 percent). The 302 students received $5 million in funds again in 1985, but no new grants were awarded because no additional money was available. Appropriations for 1986 were cut to $3 million, which was used to cover the existing students (albeit at reduced levels), who had been guaranteed 3 years of support. A new crop of students will be solicited in 1987, under a new $2.8 million appropriation. However, full funding for 3 years will be allocated from this 1987 appropriation to each new student accepted. The major reduction in funding coupled with the new policy of "forward funding" means that support will be available to substantially fewer students.

POSTDOCTORAL

The Agricultural Research Service (ARS) of USDA also initiated a competitive postdoctoral program in 1984 that supported 21 people for 1–2 years working on specific projects at ARS laboratories. The number of award recipients increased in 1985 and 1986 to 50 and 100, respectively. The 1986 appropriation for the program was $4 million, with about half of the fellowships supporting researchers in agricultural biotechnology. ARS fellowships pay $26,000–$31,000 per year, compared with NIH postdoctoral appointments, which pay $16,000–$30,000 per year, depending on the individual's experience. The ARS program is an important incentive in attracting young scientists to agricultural research.

LAND-GRANT SYSTEM

Most research in plant biology is conducted at land-grant universities, which also support 80 percent of the nation's plant biology faculty and graduate students. As training centers in plant biology, land-grant universities must continually update programs to reflect trends in biotechnology and must equip students with the knowledge to apply biotechnology to important problems in agriculture.

THE IMPORTANCE OF PEER REVIEW

The importance of peer-reviewed, competitively awarded federal grants in supporting talented pre- and postdoctoral students has been demonstrated. Follow-up studies on recipients of NIH grants show that they outperform nonrecipients in their subsequent careers in biomedical research (IOM, 1983, 1985). Yet only 60 percent of postdoctoral researchers in plant biology receive federal support (Anderson, 1984), compared with 85 percent of biomedical postdoctoral researchers (IOM, 1983).

Furthermore, plant science is underfunded in proportion to the number of students in the field. The $98 million in federal funds used to support plant biology research was only 4 percent of federal funds for life sciences in 1982, although plant biology graduate students accounted for 12 percent of all graduate students in life sciences and 17 percent of the doctorates awarded (Anderson, 1984).

Given the small number of postdoctoral fellowships awarded by USDA and the fact that they are restricted to ongoing research programs at ARS laboratories, it is not surprising that three-quarters of the new Ph.D.s in agricultural science do not plan on any postdoctoral training (NRC, 1983). This situation is particularly discouraging for biotechnology, which relies more than many other agricultural disciplines on basic research. An intensified national effort is needed to identify promising graduate students and postdoctoral researchers for agricultural biotechnology and award them peer-reviewed, competitive grants through USDA programs.

Programs at the National Science Foundation

PREDOCTORAL

NSF has supported peer-reviewed, competitive predoctoral fellowships in the basic sciences and mathematics since 1952. About 450–540 new 3-year awards are made each year from an annual appropriation of approximately $27 million; 25–35 percent of the awards are in the biological and biomedical sciences.

POSTDOCTORAL

NSF has peer-reviewed, competitive programs that fund post-doctoral fellows in plant biology, environmental sciences, and mathematics. The plant biology postdoctoral fellowships attempt to foster retraining for an interdisciplinary approach to plant science. Initiated in 1983 at an annual cost of about $1.2 million, these fellowships are awarded to about 20 recent Ph.D.s each year to encourage them to explore a new research direction in plant science—for example, to help a bacterial molecular biologist switch to plant molecular biology or a plant tissue culturist to investigate plant biochemistry. NSF's environmental biology fellowship program began in 1984 and supports about 20 people each year at a cost of about $1 million. Similarly, the mathematics postdoctoral fellowships have supported about 30 people each year since 1979 at a cost of about $1.5 million per year. In addition, NSF funds North Atlantic Treaty Organization (NATO) fellowships for post-doctoral study in science and engineering by U.S. citizens working in NATO countries and NATO-affiliated countries. NSF awards about 50 NATO fellowships per year at a cost of around $1 million.

SUMMER COURSES

Since 1981, NSF has funded a summer course on plant molecular biology at Cold Spring Harbor Laboratory in New York that, like the plant biology postdoctoral fellowships, aims to give scientists educated in related disciplines a foundation in this relatively new field. Sixteen people are accepted into the course each year out of about 50 applicants. Seventy-five percent of the applicants are Ph.D. scientists. The remainder are graduate students or non-Ph.D. scientists from industry.

Career Development

NSF contributes to the newly conceived Presidential Young Investigator Awards, which support the independent research of outstanding young faculty scientists nominated by their departments or deans. The purpose of this program is to help universities attract and keep outstanding young Ph.D.s who might otherwise pursue nonacademic careers. In 1984, 200 Presidential Young Investigators were named, and 100 more were appointed in 1985 and again in 1986. NSF funds the awards for 5 years at a base rate of $25,000 per year. NSF will also provide up to an additional $37,500 each year to match funds provided to the award recipient by industry. The awards are divided among the disciplines of NSF's research directorates. The number of award recipients in the biological sciences were 25 in 1984, 21 in 1985, and 10 in 1986. At least one-third of these scientists are carrying out basic research with potential relevance to agriculture. Examples are studies of growth-related peptides in livestock animals and the genetics, physiology, and biochemistry of productivity and water-use efficiency in crop plants.

Programs at the National Institutes of Health

The NIH research and training system for biomedical science is perhaps the most effective system of its kind in the world. It is by far the major source of biomedical research training support in the United States and has been instrumental in America's leading role in basic biomedical research since World War II. The extensive NIH programs contrast markedly with the very limited programs and support provided through USDA for research training.

Extramural

NIH National Research Service Awards (NRSA) support several types of extramural fellowships. NRSA predoctoral traineeships support graduate education in basic biomedical science; about 5,000 such positions were funded in 1985 at a cost of $73 million. Similarly, NRSA funded about 5,700 postdoctoral awards in basic biomedical science in 1985 from an appropriation of $145 million. NRSA awards are administered either as institutional research training grants or as individually awarded fellowships (the latter for postdoctoral researchers only). All awards are competitive and peer-reviewed.

In addition to predoctoral traineeships and entry-level postdoctoral traineeships and fellowships, NRSA extramural awards are given for Senior Postdoctoral Fellowships, Mid-Career Conversion Awards, Academic Investigator Awards, Clinical Investigator Awards, Physician Scientist Awards, Research Career Development Awards to aid young scientists setting up independent research laboratories, and Special Emphasis Research Career Awards to develop an individual's multidisciplinary capacity for research.

INTRAMURAL

NIH appoints Intramural Staff Fellows through a different peer-reviewed, competitive program. There are three categories (the number of fellows selected in 1985 is given in parentheses): (1) entry-level Staff Fellows, who have less than 3 years experience beyond the Ph.D. (318); (2) Senior Staff Fellows, with 3–6 years experience (283); and (3) Medical Staff Fellows (327), who take on both research and clinical duties. Staff fellowship positions are nontenured and may last up to 7 years. The three categories respectively allow (1) valuable training in NIH labs for junior researchers, (2) more advanced researchers to learn the latest biomedical techniques at NIH labs while bringing in their own expertise, and (3) the integration of clinical and basic medical research.

Intramural programs also sponsor Visiting Fellows (577) and Associates (228) at NIH labs. These temporary personnel exchanges involve foreign citizens and promote both retraining and the exchange of ideas between countries and laboratories.

Other Government Programs

The NRC administers a peer-reviewed, competitive Research Associate Awards program, under which scientists work as guest investigators in U.S. government laboratories. About 500 Research Associates are supported by more than 30 laboratories, including those of the National Aeronautics and Space Administration, NBS, the National Oceanic and Atmospheric Administration, and recently, NIH. ARS, however, has not participated in this program since 1977. The program attracts high-quality scientists, both recent Ph.D.s and senior investigators, who can bring stimulating ideas and new techniques to their sponsoring laboratory.

Private Support

Private support for basic research and training has generally been limited compared with federal support. Some private firms do fund fellowships, but such programs are quite limited. Recently, the privately funded McKnight Foundation broke new ground: it initiated a 10-year program to award a total of $15 million for interdisciplinary, problem-oriented university training grants in plant biology related to agriculture and a second 10-year program to award $3.5 million to outstanding young plant biologists. Both types of grants are awarded through a peer-reviewed, competitive process, and each grant lasts 3 years. The interdisciplinary grants of $300,000 per year pay mainly for pre- and postdoctoral fellowships, and similarly, the individual awards of $35,000 per year are often used to support a research fellow within the young faculty member's laboratory.

Some private firms also fund fellowships. The Federal Technology Transfer Act of 1986 (see Chapter 5) facilitates support of training positions in federal laboratories by private firms.

Conclusions

Training opportunities in biotechnology for agriculture are very limited. The USDA has recently put programs into place, but the number of trainees and the level of funding are small in contrast to the biomedical and basic research efforts of NIH and NSF (Table 4-1). Expenditures given in Table 4-1 do not include USDA Hatch Act support to the Agricultural Experiment Stations that fund predoctoral trainees as graduate research assistants. Likewise, they do not include the portions of NIH and NSF basic research grants that support pre- and postdoctoral trainees, nor NIH's intramural programs. The latter mechanisms of support are considerable. Total NSF funding of pre- and postdoctoral trainees is 3- to 12-fold higher than shown in Table 4-1 if figures for students supported by their sponsor's research grants are included.

Private support for research training is also limited and does little more than supplement government programs. Major federal

TABLE 4-1   Federal Agency Expenditures for Training Research Scientists
(millions of dollars)

Year	Agency		
	USDA[a]	NSF[b]	NIH[c]
Predoctoral programs			
1983	—	15.0	61.8
1984	5.0	20.3	61.0
1985	5.0	27.3	73.0
Postdoctoral programs			
1983	0.6	3.2	102.8
1984	0.7	4.7	105.6
1985	2.0	4.6	145.0

NOTE: The table includes all funding through specific training programs but does not include support to pre- or postdoctoral trainees provided under individual research grants.

[a] The USDA predoctoral program was initiated in 1984 and provides funds in the form of training grants to university departments. From 1981 to 1983 the 1-year postdoctoral appointments required the same civil service hiring practices used for permanent staff; beginning in 1984, special authority under the Office of Personnel Management's Schedule B has been used to expedite postdoctoral appointments.

[b] NSF predoctoral fellowships cover all scientific and engineering disciplines; postdoctoral fellowships exist under four programs only: Mathematics, the North Atlantic Treaty Organization, Environmental Biology, and Plant Biology.

[c] NIH training grants to university departments support both pre- and postdoctoral recipients. Individual NIH fellowships are only available for postdoctoral recipients.

SOURCE: Personal communications from agency program directors, 1986.

increases for training programs in agricultural biotechnology are urgently needed to stem the erosion of U.S. agricultural research capability and to meet the growing need for trained scientists. These programs must include four types of support: pre- and post-doctoral fellowships, training grants, career development awards, and retraining opportunities. They should be administered on a peer-reviewed, competitive basis.

INTERDISCIPLINARY COOPERATION

Traditional agricultural researchers are often unfamiliar with recent advances in molecular genetics and biotechnology. Conversely, molecular biologists and other scientists with expertise in modern techniques usually have little background in agricultural research. Insufficient interaction between basic and applied researchers impedes training and thus inhibits practical applications of biotechnology to agricultural production.

Human health-related research provides another route of interdisciplinary information flow into agricultural research. The development and application of biotechnology have progressed faster in health research, because of larger public and private investments. Agricultural scientists should keep in contact with the latest achievements in biomedical research, which often have direct and/or indirect significance for animal and plant research.

Agricultural scientists and research institutions need to reach out and develop new links with basic science disciplines. These new links could take a variety of forms.

Curricula. Universities can promote interdisciplinary cooperation by two complementary tactics. They must first provide a broad education for undergraduates that covers the basics of all the sciences. This should include agricultural science as well, which is often omitted from curricula in non-land-grant institutions. Conversely, colleges of agriculture should strengthen their curricula in other basic sciences. Universities must then create graduate curricula and graduation requirements that include coursework complementary to students' specialties (for example, courses in physical chemistry to understand the physiology of plant stress).

Training Grants. Peer-reviewed, competitive training grants for research areas spanning several disciplines are another way to effectively educate pre- and postdoctoral students and at the same time promote interdisciplinary cooperation. These grants provide stipends for students and may also cover the costs of equipment and research. By bringing common goals to several different fields of research, such training grants can encourage young scientists to creatively apply ideas and methods from complementary disciplines.

Career Development. Similarly, career development awards for young, independent faculty contribute both to the advancement of research and to the education of students. The NIH, NSF, and NRC fellowship and career development programs show the effectiveness of federally supported peer-reviewed, competitive awards for educating and training researchers.

Retraining. Faculty sabbaticals and senior postdoctoral appointments that cross traditional disciplinary lines are very important avenues for the exchange of ideas and personnel retraining. Providing established researchers with opportunities to learn new biotechnological methods capitalizes on their existing expertise in agricultural systems. Furthermore, these established agricultural scientists will be instrumental in educating the next generation of researchers. Their adoption of biotechnology will allow them to teach students about agricultural science in a way that integrates classical and modern approaches.

Biotechnology relies on large-scale team approaches, orchestrated both within and among laboratory groups. Thus, interdisciplinary cooperation is needed for the growth of agricultural biotechnology and its application to real-world problems. This is true not only for industrial R&D but also for solving complex problems in the underlying biological sciences. For example, communication and cross-training among laboratories studying entomology, neurochemistry, and molecular biology are essential for a modern approach to pest control through biochemical modification of insect behavior. These types of interdisciplinary projects must be supported with new sources of funding and new rewards. They also require curricula and educational programs that give collaborating researchers an understanding of each other's fields.

RECOMMENDATIONS

Scientists, administrators, faculty, and policymakers in all sectors should be aware of the importance of state-of-the-art education and training to the future development of agricultural biotechnology. Specifically, the committee makes the following recommendations.

INCREASED FEDERAL SUPPORT FOR TRAINING

Major increases in federal support for training programs are urgently needed to provide a high-quality research capability that ensures the future of U.S. agriculture and meets the growing need for scientists trained in agricultural biotechnology. Four types of programs must be supported: pre- and postdoctoral fellowships, training grants, career development awards, and retraining opportunities. These approaches, used successfully in the biomedical sciences, have put the United States in the forefront of human medical advances. These programs should be administered on a peer-reviewed, competitive basis. USDA should support at least 400 postdoctoral positions at universities and within the ARS, which represents a quadrupling of the present number, and maintain strong support for graduate-level training.

INCREASED RETRAINING PROGRAMS

For the short term, highest priority should go to increasing the retraining opportunities available to university faculty and federal scientists to update their background knowledge and provide them with laboratory experience using the tools of biotechnology. This retraining will expand the abilities of researchers experienced in agricultural disciplines. USDA should take the lead in administering a program to supply at least 150 retraining opportunities a year for 5 years, starting in FY89.

5
Technology Transfer

INTRODUCTION

The goal of technology transfer has always been implicit in U.S. science policy: Federally funded research should benefit the public, and such benefit includes the development and transfer of technologies from public laboratories to the private sector.

Yet what in theory appears to be a simple process of translating basic research discoveries into social benefits and commercial applications is in reality a complex set of interactions involving many types of people and institutions. Technology transfer involves the flow of information between basic and applied research and the subsequent transfer of products of research to dispensers and ultimate users. This chapter examines several of the mechanisms that facilitate the exchange of information in technology transfer and recent developments in relationships among universities, industries, and government. It also looks at how patent policies are changing patterns of technology transfer in agriculture.

The Economic Dimension

Technology transfer is propelled by the potential benefits derived from using and adapting a research discovery. Economic incentives spur people to improve and transfer technology. Industry will not develop and market nor will farmers adopt new technologies without clear, perceived payoffs. However, improved technologies are often blamed for the current huge agricultural surpluses. Quite the contrary, the causes of surplus agricultural commodities lie elsewhere.

When adopting new technologies can increase sales and profits by reducing costs, farmers will choose them to improve their competitive position. In the new global marketplace for agricultural trade, American farmers are competing with other producers throughout the world. Technological improvements and efficiency are critical components in this competition. It is clearly in the public's interest to ensure that the U.S. agricultural research system, including the many interconnections that promote technology transfer in agriculture, are in place and fully operational. National policies must facilitate the use of new technologies in agriculture.

The seed industry is an example of the interrelation of funding research, institutional roles, technology transfer, and productivity. Historically, breeding improvements in openly pollinated grain crops, as opposed to hybrids, were developed by public institutions. Breeding programs to locate and incorporate pest resistance and other yield-enhancing traits are a long-term research investment. New traits from the publicly supported breeding programs were made openly available to commercial breeders for seed production. Recently, public funding for this basic breeding work has been reduced and private companies have become active. Yet are U.S. farmers prepared to pay the long-term costs of breeding work in the price of seed? The changing patterns in technology development and transfer could lead to loss of productivity growth in varietal performance, higher food costs, and loss of competitiveness in world trade. This then brings us to the issue of public/private cooperative development and the transfer and adoption of new technology.

UNIVERSITY, INDUSTRY, AND GOVERNMENT INTERACTIONS

Challenges to U.S. technological superiority have appeared across a range of industries. In part, this situation results not from a lack of technological expertise but from inadequate technology transfer and product and process development based on results of fundamental research. Transferring technology between academic and industry scientists in the biological sciences used to occur informally and by chance as scientists conversed at meetings. However, recent breakthroughs in molecular biology and biotechnology and their potential commercial implications have led to more formal and aggressive efforts. Technology transfer is important in the interests of industrial competition. The shift has been toward the promotion of collaborative research relationships between publicly supported scientists in universities and federal laboratories and those in the private sector. Laws such as the Stevenson-Wydler Technology Innovation Act of 1980 (P.L. 96-480), the Small Business Innovation Development Act of 1982 (P.L. 97-219), the Federal Technology Transfer Act of 1986 (P.L. 99-502), and recent proposals to liberalize patent policies have strengthened the emphasis on technology transfer in the nation's science agencies.

The Stevenson-Wydler Technology Innovation Act of 1980 designated the U.S. Department of Commerce as a lead agency for federal technology transfer, with additional support coming from the National Science Foundation (NSF) and the federal laboratories. Efforts were to be coordinated by a number of offices and centers for industrial technology, research, and applications. These were designed to promote the use of results of federally funded R&D by the private sector as well as state and local governments. The Federal Technology Transfer Act of 1986 amended the Stevenson-Wydler Act by authorizing government-operated laboratories to enter into cooperative research agreements and by providing incentives for commercializing federal patents. The Small Business Innovation Development Act strengthened the role of small, innovative firms in federally funded R&D by requiring federal agencies with R&D budgets of $100 million or more to set aside a percentage of their funds to support R&D done by small businesses.

Universities as well as state and federal agencies are expanding their relationships with the private sector as they explore ways to increase scientific communication and the flow of technology. Breakthroughs in biotechnology have greatly shortened the time between basic discoveries and product development. Opportunities to establish links between basic and applied research programs, and financial incentives including consultancies, patent agreements, and grants and contracts from industry are having a positive effect on technology transfer. The following section describes some of these relationships between university and government research and industrial development in agricultural biotechnology.

Research Relationships in Technology Transfer

With the growth of biotechnology programs in the early 1980s, universities and industry competed for scientists with skills in biotechnology research. This competition has led, in part, to new relationships between university scientists and industry. These relationships try to address the needs of both groups, and they survive as long as both benefit. Although most of the university–industry–government links have counterparts in engineering and related scientific disciplines, biologists are relatively new to such collaborative arrangements.

Five general types of alliances are evolving: (1) programs that are part of general university efforts, which normally include graduate student training and publication of scientific findings; (2) projects that have a defined application, which may include a proprietary interest in achieving certain results; (3) programs that are directed to commercializing faculty research; (4) programs that operate outside the university to aid clients; and (5) free-standing institutes linked to several universities (Government–University–Industry Research Roundtable, 1986).

These diverse approaches reflect the fact that universities encompass a diverse set of roles and interests. Thus, universities are evolving and testing a variety of structures for their alliances with industry. What works for one alliance may not suit another. Clearly, there is a need for a range of approaches.

Similarly, universities and companies must address problems of conflicts of interest and ownership of intellectual property in the context of their relationship. Solutions will depend on their individual situations and needs. It is up to each side to protect its own interests.

Most of the mechanisms used to develop mutually beneficial alliances among universities, industries, and government include one or more of the following.

CONSULTANCIES

University faculty have traditionally consulted with industry on an individual basis, contributing expertise in science or to solving a particular problem. This exchange of information between academic scientists focusing on basic research and industrial scientists concerned with product development is a major means of technology transfer. Consultancies are increasingly common, particularly in biotechnology, as start-up companies and established chemical and drug houses mount research programs in this area. In fact, it is difficult to find a prominent university molecular biologist who does not consult to the biotechnology industry.

There are legal concerns when consultancies are extended to federal employees. For example, is it proper for an individual on the federal payroll to serve one person, group, or company to the exclusion of others? Guidelines on federal employee consultancies should consider three concerns: conflict of interest, favoritism, and mutual benefit.

These guidelines govern the current policy of the Agricultural Research Service (ARS) on consultancies between its scientists and the private sector. However, the number and scope of current arrangements are limited. On the other hand, the National Bureau of Standards (NBS) has long played a primary role as a consultant to and collaborator with industry. Scientists at the NBS may consult to industry as representatives of NBS if the subject matter falls within the bureau's mission. If the expertise required is not related to their jobs, these scientists may consult as private individuals. Recently, the National Institutes of Health (NIH) also instituted flexible policies on consultancies between their scientists and industry. NIH scientists may use their general knowledge and expertise to consult for particular individuals, companies, and institutions. Ongoing NIH research results, however, may only be disseminated through nonexclusive channels such as open lectures and conferences. The open policies of NIH and NBS have encouraged the transfer of technology from government-funded basic research into practical applications that benefit society as well as their industrial developers.

Consultancies assist scientific advancement beyond the remunerative benefits to individuals, corporations, and government organizations. Consultancies can foster technology transfer, and when they lead to more formal university–industry–government agreements or consortia, they usually provide funding and training opportunities for students and benefit research through interdisciplinary research collaborations.

EDUCATION AND TRAINING

Education and training arrangements exist on several levels. Companies give "student gifts" that pay stipends for undergraduate, graduate, or postdoctoral positions, sometimes to be used by a university department at its discretion, sometimes earmarked for an individual professor, or sometimes for training in an area important to the company. Another type of arrangement is the "industrial affiliate." Companies send their scientists to universities as affiliates, to learn about departmental programs, to meet with faculty and students, to perhaps have access to findings prior to publication, and to possibly identify promising students as future employees. Affiliate programs benefit universities by fostering consulting arrangements and research contracts and by teaching universities about the needs, especially student training needs, of industrial research laboratories. In some cases they also provide significant funding for stipends and the enhancement or expansion of graduate programs.

GRANTS AND CONTRACTS

Grants and contracts between universities and industry range from general grants for basic research to specific contracts for defined projects. The sizes of such grants and contracts vary, ranging from a few thousand dollars to much larger sums as part of long-term industry–university arrangements. The smaller contracts and grants to State Agricultural Experiment Stations (SAESs), however, can be reasonably significant amounts (see Table 3.5). For example, support to the California, Texas, and Florida SAESs from industry grants and contracts in 1984 totaled $9.0 million, $6.6 million, and $4.7 million, respectively.

A number of large biotechnology grants have recently been awarded by industries to university research institutes or laboratories. Examples include the Hoechst Department of Molecular Biology at Massachusetts General Hospital, initiated with a $70 million, 10-year award, the Dupont-supported Department of Genetics at Harvard Medical School, and Monsanto's $23.5 million, 5-year grant to the Department of Medicine at Washington University. Such large grants promote multidisciplinary work within departments, a necessary component of biotechnology research. These arrangements involve more than a simple transfer of funds: The company and the university must define their roles in the R&D efforts. This is necessary in order to maintain the integrity of both academic and industrial values. The former—public knowledge, publication, and peer evaluation—can conflict with the latter—proprietary knowledge and products. Linkage institutions (discussed in this chapter) can mediate these potential conflicts and establish some degree of compatibility between university and industrial systems. Both partners can gain an appreciation of their respective values, capabilities, and constraints (Omenn, 1982a).

CONSORTIA AND RESEARCH PARKS

Consortia combine the strengths of several companies with a university, or alternatively, unite the strengths of several universities. Consortia serve as centers of excellence, technology transfer, and training. Industrial research parks, another innovation, can breed small companies linked to a university. Several state and local government groups are involved in creating incubator centers that include expensive facilities and equipment as shared services to attract biotechnology companies to their area.

TECHNICAL DEVELOPMENT OFFICES

Universities and state and federal government agencies seeking to promote the development and licensing of patentable inventions have created programs to encourage technical development. These programs range from staff to assist scientists filing for patents to entrepreneurial efforts that control licenses and commercialize patented inventions. (University and government patenting activity is discussed in more detail later in this chapter.) Relatively few resources have been allocated to technology transfer by federal laboratories. The Federal Technology Transfer Act of 1986 should stimulate efforts in this regard.

ENTREPRENEURIAL COMPANIES

A significant number of scientists leave university or government posts to work for companies or to start their own companies. A recent survey revealed that one-third of the founders of responding biotechnology firms previously had been associated with universities (Magrath, 1985). Examples include Agracetus, BioTechnica, Calgene, Damon Biotech, Integrated Genetics, and Molecular Genetics. Some faculty work part-time in industry or have equity ownership.

Alliances Related to Agriculture

Of the many alliances established among universities and corporations, and in some instances government agencies, several focus on agriculturally related research. The following examples illustrate the diversity of approaches and the levels of funding involved in these alliances.

CORNELL UNIVERSITY BIOTECHNOLOGY PROGRAM

The Cornell program began in 1982 with funds from New York State and a 6-year commitment from three companies: Union Carbide, Eastman Kodak, and General Foods. In 1986, the program was designated a Center of Excellence in Biotechnology by the Army Research Office under the University Research Initiative Program. This status provided additional financial support. Annual support through the program amounted to 10–15 percent of the total investment in biotechnology research at Cornell, which was approximately $20 million in 1985.

Cornell faculty compete for funding from the consortia by submitting research proposals to the biotechnology program. Six representatives of the university and three from the participating companies review the proposals, and award grants of about $50,000 per year. In addition, the program hosts resident industrial scientists at Cornell and sponsors symposia and workshops, bringing together university researchers, corporate vice presidents, and scientists from the sponsoring companies. Central support facilities, such as for DNA synthesis, protein sequencing, and so forth, are also operated by the program.

The key feature of the Cornell biotechnology program is its emphasis on interdisciplinary research. Such research suits the program's broad agenda: exploration of the molecular aspects of cell biology and genetics as they apply to agricultural problems. Topics range from basic research on gene regulation and manipulation to applied problems such as scaling up cell culture systems for industrial production. The program's ultimate goals are to increase agricultural productivity within the next 5–10 years through improved livestock species, animal vaccines, and plants resistant to pathogens and environmental stresses, and to use cell products for special chemicals, toxic waste control, and as sources of protein.

Another important aspect of the program is an economic development committee, which studies product marketing. Cornell owns all patents on inventions coming out of the biotechnology program. Participating companies are not guaranteed exclusive licenses, but once they have acquired a license, they do not pay royalties to the university. The rationale for this, as well as for the companies' use of unpatentable information, is that Cornell receives its share from the companies' initial support.

PITTSBURGH PLATE GLASS / SCRIPPS CLINIC

Pittsburgh Plate Glass (PPG), which has been in the agrichemicals business since the early 1940s, entered into a joint venture in 1985 with the Department of Molecular Biology of the Research Institute at Scripps Clinic in La Jolla, CA. The 15-year agreement provides $2 million a year for basic biotechnology research in plant science, with annual increases for a total of $50 million. PPG has put up an additional $10 million for a new building, which belongs to Scripps and houses more than 100 researchers. These researchers will all be employees of Scripps; their salaries and basic research budgets will be provided by federal research grants, for which they compete. PPG's money, which amounts to 10 percent of the department's $20 million operating budget, will be used to buy new research equipment. In return, PPG is assigned rights for developing anything patented by Scripps involving agrichemicals, plant species, or microbial strains. PPG both pays for and decides what to patent. The PPG/Scripps arrangement parallels one established in 1982 between Johnson & Johnson and the Scripps Department of Molecular Biology for health-related research.

MICHIGAN BIOTECHNOLOGY INSTITUTE

The Michigan Biotechnology Institute (MBI) is a nonprofit corporation dedicated to the commercialization of biotechnology and the development of renewable resource-based business opportunities in the Midwest. The institute emphasizes industrial applications of biological sciences, focusing on research and development of new products and processes, technology transfer, and collaboration among industrial, university, and national laboratories. Specific areas of interest include industrial enzyme technology, biomaterials and fermentation technology, and waste treatment biotechnology.

MBI was created in 1983 and initial funding was provided by the state—$6 million through 1987. As of August 1986, MBI had raised an additional $33 million from private sources and state loans. The institute employs 50 business and scientific personnel.

The MBI business division handles commercial market analysis, fund raising, patents, contracts for R&D with industry and government, and the coordination of public relations and educational programs. The research division consists of a scientific staff, primarily biologists and engineers, who may hold joint appointments with Michigan State University or other universities. There are also adjunct scientists—full-time university professors who work for MBI as consultants or as professors for the training programs, and project interns and trainees, who are graduate students and postdoctoral fellows.

MBI's goal is to facilitate interaction between universities and industry that will lead to economic development. By positioning itself as a nonprofit corporation between academia and commercial companies, MBI links these two groups. It supports single-discipline, problem-focused research done in universities, thereby helping to generate patentable ideas. It then directs this knowledge, through a multidisciplinary approach with an emphasis on

R&D and economic analysis, into proprietary processing and product application for industry. Industry performs the final task in the discovery–application–commercialization scheme by marketing products and processes.

NORTH CAROLINA BIOTECHNOLOGY CENTER

This private nonprofit corporation was established in 1981 as the nation's first state-sponsored initiative in biotechnology. It is largely funded by the state of North Carolina, which for the 1985–1987 biennium appropriated $14.2 million to the center. The center promotes statewide R&D in biotechnology by initiating, sponsoring, and funding research, university–industry collaboration, commercial ventures, meetings, and program activities. The center, located in North Carolina's Research Triangle Park, is not itself a site for research.

The center encourages research and activities that are multidisciplinary and multi-institutional, that lead to university–industry collaboration and technology transfer, and that will result in useful products. The center catalyzes interactions among parties involved in biotechnology development, fosters development of biotechnology industries within the state, funds research faculty recruitment and facilities development at the universities, and provides public education about biotechnology. Current programs include the Monoclonal Lymphocyte Technology Center, the Biomolecular Engineering and Materials Application Center, the Bioelectronics Advisory Committee, the Bioprocess Engineering Feasibility Study Committee, Visiting Industrial Scientists and Engineers at North Carolina Universities, the Marine Biotechnology Advisory Committee, the Program in Public Information and Education on Biotechnology, and the Triangle Universities Consortium for Research and Education in Plant Molecular Biology. In FY85–86, the Competitive Grants Program awarded $833,000 to 44 projects, and the Industrial and University Development Grants Program awarded $3.8 million for various biotechnology activities, research, and development statewide.

NEW JERSEY CENTER FOR ADVANCED BIOTECHNOLOGY AND MEDICINE

This program is one of several state-supported advanced technology centers recently created in New Jersey with the aim of attracting the best scientists and providing an environment for basic research that can lead to industrial development and subsequent economic strength in the state. The state will provide the center with 50 percent of its research and salary budget, the other 50 percent is to be covered by competitive federal grants and industrial participation once the center is fully operational. The state appropriations for the center's operating budget for FY85, FY86, and FY87 are $1.3, $1.5, and $3.2 million, respectively. A building project is being financed by general obligation and revenue bonds totaling $35 million for the construction of the center and two satellite facilities.

The Center for Advanced Biotechnology and Medicine has a scientific advisory board of senior faculty and prominent outside scientists who are helping to recruit the scientists to head its 18 research teams that will focus on human molecular biology. The center will be located on Rutgers' Busch Campus and jointly operated by Rutgers University and the University of Medicine and Dentistry of New Jersey (UMDNJ)–Robert Wood Johnson Medical School. The clinical research unit is located at the Robert Wood Johnson Hospital. The Waksman Institute of Microbiology focuses on cell fermentation processes and technologies; the rest of the Waksman Institute is redirecting its research into two areas important to biotechnology: regulation of gene expression and biomolecular structure, both of which will include basic research on plants and animals. The Waksman Institute will also have greenhouse and field space. Thus, the many programs at the center, the Waksman Institute, and the UMDNJ–Robert Wood Johnson Medical School form a concentration of biotechnology research in New Jersey. Plans for an Advanced Technology Center for Molecular Biology in Agriculture are being explored. Its research on plants and animals would complement the Center for Advanced Biotechnology and Medicine's research on human molecular biology.

UNIVERSITY OF CALIFORNIA BIOTECHNOLOGY RESEARCH
AND EDUCATION PROGRAM

This statewide competitive grants program was begun in 1985 and has a current state appropriation of $1.5 million. The program has an indefinite authorization; the university will request increases in the budget in coming years. Each of the nine campuses of the University of California may submit one research proposal per year. A committee of representatives from each campus makes three to four awards per year on the basis of faculty reviews of the proposals. The awards are made as training grants of $200,000–$300,000 covering a 3-year period. There is also an advisory committee composed of representatives from the biotechnology industry and agencies outside universities, which recommends directions for funding.

PLANT GENE EXPRESSION CENTER

The ARS and the Regents of the University of California are cooperating in the establishment of a research center at Albany, California. The program will study the complex biology of plant genes, the control of their expression, and the biochemical steps and developmental mechanisms responsible for the quality and productivity characteristics of plants. A mandate of the Plant Gene Expression Center is to strengthen the research relationships among ARS, university, and other scientists pursuing new technologies to improve crop plants.

This center is a federally funded research facility that will have a core scientific staff of 10 senior researchers. Two of the senior researchers will hold full faculty positions at the nearby University of California at Berkeley, and the other 8 will be hired under procedures that qualify them for adjunct faculty status. This arrangement gives the state university system and the federal ARS system a collaborative role and responsibility in developing and maintaining the scientific quality of the center. By design, this is a long-term commitment to a basic research program by both the federal and state cooperators, and it combines the approaches of federal research teams and university principal investigators to scientific research.

The research facility at Albany is planned to house, through direct employment or other arrangements, about 70 scientists, postdoctoral researchers, and other research associates besides the support staff. Operating funds will come from the ARS. The budget is projected to be $2.5 million in FY87. When fully operational, the center will have an annual budget of about $6 million. In addition, the university-associated researchers are eligible to apply to federal granting agencies for additional research funds.

Implications of Alliances and Research Relationships

Basic discoveries in biotechnology can often be translated into commercial applications. Thus, industry seeks ties to university and government laboratories and vice versa. However, the relationships described earlier and other programs in biotechnology involving university–industry–government alliances are still too new to be judged on their effectiveness at promoting agricultural biotechnology. At this stage, therefore, it is best to regard them as models that illustrate the diversity of approaches available to promote interdisciplinary research and cooperation among industry, universities, and government.

Each research sector performs complementary tasks and seeks to gain something through its relationships. Industry gives funds for basic research, which universities typically perform more efficiently. Likewise, industry supports some training of scientists in university laboratories. In return, industry gets direct access to the results of research programs: "know-how," "show-how," and immediate practical applications. Furthermore, industry is able to hire new graduates trained in the areas of expertise it seeks.

The universities, in turn, provide a strong environment for basic research and a training ground for scientists. Their grants and contracts with industry provide money beyond what they can obtain from state and federal governments. The universities may also profit from their role in developing intellectual property and their tie-in to applied problems.

Finally, government participation strengthens the foundation of research and training programs. Government involvement provides a center of activity, which attracts industrial development and promotes economic growth, which in turn benefits the entire nation.

University–industry research relationships supported between 16 and 24 percent of university biotechnology R&D in 1984 (Blumenthal et al., 1986). Although this is a far higher proportion than industry's overall contribution to universities, these funds represented less than 10 percent of the R&D budgets of most firms. This proportion of industrial investment in university research probably will not increase further, and will most likely decline as a direct consequence of successful technology transfer. As companies identify more potential products, they will shift their financial support to conduct more research in-house, particularly applied research that leads to patents. Nevertheless, industry will still look to universities for advances in fundamental research.

Many scientific advances that made biotechnology possible came out of basic research funded by the federal government. Other nations have also made valuable contributions. University–industry research relationships and their commercial consequences clearly show the practical value of long-range government funding to universities. Industrial alliances now offer new gains for universities: increased income from grants, contracts, and, potentially, patent royalties and licenses; program expansion; and student opportunities.

Potential risks, however, stem from the dichotomy in academic and industrial value systems—public versus proprietary knowledge and products. These risks include constraints on the communication of research, bypassing peer review of grants (Omenn, 1982a), tracking of students onto industrially oriented projects, faculty conflicts of interest, and some tendency of industry to award short-term grants or to favor applied over basic research goals (Blumenthal et al., 1986). Industry may also try to dictate the direction of research, or seek out and fund only those projects close to fruition. However, universities can, by judicious bargaining, put their interests foremost to minimize such risks. The linkage institutions discussed in this chapter are important in mediating successful university–industry collaboration.

Industry's funding is not great compared with government funding of basic research. Moreover, industry rarely funds whole departments or even whole laboratories—industry's grants usually leverage existing facilities and expertise, as shown by Cornell's program. When industry does initiate construction of university laboratories or hiring of new faculty, as in the PPG/Scripps arrangement, industry's money provides only a small fraction of the total operating budget. Most of a department's expenses and faculty salaries come from other sources, including state and federal government grants. For instance, industry funds only 10 percent of both the Cornell and Scripps biotechnology programs. Therefore, industry cannot be expected to compensate for any reduction in federal funding. The continued health of research efforts at universities remains highly dependent on federal and state governments as major sources of support.

MERGING BIOTECHNOLOGY INTO AGRICULTURE

Identifying problems that biotechnology can address and introducing new products from biotechnology into agricultural practice are two important steps in technology transfer. An additional aspect is the ultimate effect of the technology. A new technology can have three orders of effects (Kiesler, 1986):

The first is the intended technical effects—the planned improvements . . . in new technology. The second is the transient effects—the very important organizational adjustments made when a technology is introduced but that eventually disappear. The third is the unintended social effects—the permanent changes in the way social and work activities are organized.

Biotechnology applications to agriculture can be expected to have the same orders of effects. The colleges of agriculture in the land-grant university system and the agricultural extension system can be expected to help implement the first effect, the intended improvements. In addition, these institutions will be strongly influenced by the third effect, the unintended and permanent social effects. These institutions can play an important role in recognizing these effects and helping individuals cope with them. The second effect, transient adjustments, is now taking place in terms of questions on regulation and public concerns over environmental considerations in field testing.

Land-Grant Universities

Some land-grant universities have recently initiated programs to support biotechnology by measures ranging from creating biotechnology institutes or centers to reallocating funds for biotechnology research and reclassifying faculty positions to recruit molecular biologists. At present most funding for these initiatives is public (Buttel et al., 1985). However, successful programs can be expected to draw industry support to their states.

Industry has a definite role in supporting research at land-grant universities, particularly in the area of biotechnology. Plant breeding departments are the major focus of some biotechnology initiatives at land-grant universities. In addition, these universities are slowly expanding their research in agricultural science to include more basic aspects of molecular and cell biology. However, the line between basic and applied research is often blurred in biotechnology, especially as it applies to agriculture, where researchers have traditionally tackled both basic and applied aspects of a problem.

Biotechnology centers at land-grant universities have several critical functions in technology transfer. They facilitate the exchange of information and ideas between scientists working on applied aspects of plants and animals and their university colleagues studying basic aspects of molecular and cell biology, biochemistry, and related disciplines. Centers attract students from traditional agricultural departments who seek minors study in biotechnology. Centers inform agricultural scientists of future research agendas in biotechnology (Buttel et al., 1985). They also facilitate vertical and horizontal integration of research.

Non-land-grant institutions such as Massachusetts General Hospital and Harvard University have recently begun programs for basic research in plant science, previously an area of limited interest outside the land-grant universities. Some large agrichemical companies have moved in to fund this research. The implication is that the land-grant universities are doing insufficient research to support applications desired by industry (Buttel et al., 1985).

The Division of Agriculture Committee on Biotechnology has published guidelines for university–industry research contracts (National Association of State Universities and Land-Grant Colleges [NASULGC], 1984) designed to promote productive and equitable interactions between land-grant universities and private industry. All universities are authorized to confer exclusive licenses to companies under the Patent Act (P.L. 96-517), although actual patent ownership may be transferred only to organizations whose mission is to transfer technology (e.g., the Research Corporation; see the section on Patents and Universities). Although full title to federally funded inventions cannot be transferred to commercial firms, public policy can encourage land-grant universities to confer exclusive licenses to private companies able to translate their discoveries into commercial products. A successful example of this is the cancer drug cisplatin, developed with National Cancer Institute funding at Michigan State University, and subsequently licensed exclusively to the drug company Bristol Meyers.

Cooperative State Extension Service

Extension is an essential part of the knowledge development, applied research, and technology transfer continuum. Technology transfer in agriculture usually carries the added challenge of adapting research developments to a range of different regional requirements. Uncontrollable factors such as climate, topography, and a host of other ecological variables dictate which agricultural innovations ultimately succeed. This fact is a major reason why agricultural scientists have maintained close communication with the users of agricultural technology. The agricultural extension system serves an important function in this communication link, disseminating research knowledge, helping to adapt that knowledge to regional problems, and reporting back the needs of the user groups.

The Cooperative State Extension Service (CES) was established in 1914 with the charge to transmit land-grant university and USDA-generated knowledge to rural people. A partnership of federal, state, and local governments carries out this mission. Roughly 37 percent of the support comes from the federal government.

As an agent of technology transfer, CES must help bring the achievements of researchers into the whole agricultural system. Here the frequent and informal contacts that occur between agricultural research scientists and the 3,000 CES specialists, who are mostly housed in the same departments at land-grant universities and the 9,000 county- and campus-based farm advisors, are crucial. Extension agents must be highly integrated with the research establishment to enable them to communicate a level of knowledge and technical skills exceeding that of the user groups for whom they provide information and training. Without this close interaction and communication with research scientists, their influence as extension agents is greatly diminished. As land-grant institutions move toward basic research, more responsibility for applied research may fall to specialists in the CES. Accordingly, they should expand their role to include some applied agricultural research, working closely with university faculty to develop the site-specific information needed in extension programs.

CES must work with biotechnology as it is developed to the stage of implementation. To do this effectively, CES must hire and train sufficient personnel with requisite expertise in biotechnology to serve as a feedback mechanism to basic researchers and to help target biotechnology research to the needs of the agricultural community. CES must be able to help different-sized farming operations and other segments of the agricultural community adopt biotechnologies and adapt them to their needs. CES should also play a role in helping the agricultural community cope with the social and economic implications of biotechnologies. These are logical extensions of CES's traditional role in community development and technology transfer.

In addition, CES must increase its contacts with the private sector, in order to evaluate new products for farmers and monitor their use. Complex agricultural technologies have spawned company marketing representatives and private consulting firms that instruct or provide specialized services for the agricultural community. They are agents of technology transfer, but they serve only those clients who can pay. Publicly supported extension agents must continue to serve the agricultural community. CES can be an arbiter of scientifically and economically sound agricultural practices. Furthermore, CES is an important source of information on environmental issues and environmentally sound agricultural practices.

Regulation and Field Testing

Progress toward field and environmental testing of genetically engineered products has been extremely slow, having relied on public agencies in their traditional research and regulatory capacity. The public debate over regulating field testing research of recombinant organisms has been going on for more than 3 years.

Controversy and confusion among the federal regulatory agencies has led to uncertainty within the biotechnology research community and industry. This has resulted in signficant delays in any field research on potential agricultural biotechnology products. Several companies have had their field testing plans delayed for a year or more, as the federal government attempts to decide which agencies are to handle field testing requests and what regulatory review procedures should be used. These delays have resulted in corresponding delays in acquiring research information from field and environmental testing, as well as in the potential introduction of beneficial products for agriculture.

The inability to conduct initial, small-scale field research with genetically engineered products is a major barrier limiting the development of biotechnology products for crop agriculture. Although laboratory tests can be devised to assess many potential benefits and possible risks associated with the use of a genetically engineered product, ultimately there is no substitute for field or environmental testing. The practical benefits and advantages of a genetically engineered product and any needed modification in the way it will be used can only be determined under conditions that parallel its potential commercial use. Such field trials not only test the effectiveness of a new product of biotechnology but also can reveal problems that warrant redesign, cautions, or regulation in its use.

There has been progress toward implementing a coherent federal regulatory program (Office of Science and Technology Policy, June 26, 1986), but widespread public confusion exists over what is being done and what still needs to be done to adequately test and regulate genetically engineered organisms. Although many of the environmental concerns raised in the course of public debates may be valid and may require scientific attention, the concern over disastrous risks associated with products of agricultural biotechnology is based largely on conjecture. Valid environmental concerns, however, must be considered. The federal government for the first time is imposing significant regulatory requirements for products with no known hazards. Moreover, it is applying these requirements at the research stage to regulate proposed research in limited-size field plots based on laboratory greenhouse-tested materials. Under these circumstances, it is incumbent on the public sector to provide an option for these initial field tests to be undertaken in a manner that permits research and product development without undue delays, while ensuring the public safety.

A decade ago the public sector had to play a major role to facilitate laboratory research on recombinant DNA in a manner judged to be safe. In response, NIH established the Recombinant DNA Advisory Committee. Now it is necessary for the public sector to again play a role to facilitate field research in a manner judged to be safe.

Over the past few years the state and federal partnership in agriculture has implemented a National Biological Impact Assessment Program (NBIAP), which recognizes the role of existing agricultural research and extension capabilities in assuring the safety of biotechnological research (NASULGC, 1986). This program is based on the precepts that research using recombinant DNA methods is not fundamentally different from other genetic research, and that the safety record of the existing framework of more than 3,000 field and laboratory locations across the United States shows that they can provide an effective, decentralized scientific capability for research involving both recombinant DNA and other methodologies. The NBIAP operates under current and emerging guidelines and public policy statements issued by the federal government on research involving recombinant DNA molecules. More specifically, it is a workable and responsible system that allows USDA to promote biotechnology research and product release into the agricultural ecosystem, while assuring that safety concerns are given appropriate attention and priority. Progress reports by the Committee on Biotechnology, Division of Agriculture of NASULGC (NASULGC, 1984, 1985, 1986) describe NBIAP.

Although NBIAP will be open to all biotechnology investigators—both public and private—an interim emphasis is needed. Initially and temporarily, the public sector should identify and establish a limited number of publicly owned, geographically isolated, and professionally managed test sites that fully meet safety needs for initial field and environmental testing. This enhanced public role in the mid-1980s is as necessary and appropriate as the Recombinant DNA Advisory Committee was when it was formed in the mid-1970s and still is.

The enhanced role proposed for the public sector will take advantage of a few selected, already existing publicly owned field stations, such as USDA experimental field stations, state and agricultural experimental field sites, and national laboratory field stations. Initially 5–10 existing sites would be selected on the basis of rigorous safety criteria such as outstanding facilities and geographic isolation. Additional significant capital expenditure should not be required for this proposal. For many such field sites long-term analytical data on soil type, climate, and other ecological factors important to monitoring environmental effects already exists. In addition, such sites have often been used for decades in controlled field trials involving pathogens and agricultural diseases.

The selected field sites should be professionally managed by an oversight committee of public sector professionals with expertise in agronomy, ecology, plant pathology, entomology, microbiology, and molecular biosciences. This committee would review proposed field research, make changes in the proposed field tests if necessary, and monitor the conduct of the tests. Public or private sector scientists desiring to use these sites would conduct the research under the observation of the site-safety officer. The site-safety officer would have overall responsibility for the safe operation and use of the test site. Costs of on-site operations would be paid by the users. Research would be conducted to gather information on environmental persistance and dispersal. These sites are proposed as an option for field tests but are not a required route for initial field testing.

To summarize, this proposed role uses existing public sites for field research and provides public professional control of research monitoring in a manner analogous to what the NIH's Recombinant DNA Advisory Committee accomplished for laboratory research. Thus, society would be protected by the collective judgment of the oversight committee, and concerns about direct private sector field research would be minimized. Without this new public role, progress toward biotechnology products for U.S. agriculture may be slow, and our nation stands to lose its current competitive advantage.

Scientific information and practical experience gained at field testing sites will help refine and streamline regulatory procedures for the public's benefit. Knowledge that ensures the public safety will help establish long-range criteria for future field testing sites that can be managed by either private or public groups. An efficient, workable, and safe regulatory system is essential to the continued progress of agricultural biotechnology in the United States. Biotechnology products are expected to provide important inputs to improve the international competitiveness of U.S. agricultural products (see, e.g., Office of Technology Assessment, 1986).

PATENTING AND LICENSING

Patents provide a means of control over the ownership of intellectual property. The owner of a patent has a form of monopoly power over his or her invention until the patent expires. Before that time, anyone else who wants to use the invention commercially must first obtain a license from the owner and in almost all cases must pay royalties. The issues of patenting and licensing are important to the progress of biotechnology because private and public investment in technology development and transfer sometimes overlap.

Although American culture generally frowns on monopolies, it makes an exception for new inventions, because the prospect of monopoly profits spurs innovation. This trade-off favors the expected long-term advantages of continued technical progress over the potential short-term gains of free access to an invention.

Technical progress depends not only on innovation but also on transferring technology from the laboratory to the marketplace. Within the private sector, technology transfer is a straightforward matter: Once an inventor is granted patent protection, he or she will be sufficiently motivated by the desire for profits to seek commercial outlets for the invention.

The public sector, however, is usually not in a position to develop and commercialize its own research. Licensing of government patents to private industry is one way to overcome this obstacle. It may seem to contradict the public interest to invest public resources in generating new technology, then restrict its use through patents and limit its benefits through licensing agreements, but such a policy is justifiable for technologies in areas in which product development involves significant capital assumption of risk. Biotechnology is such an area. Although most of the initial research in biotechnology has come from the public sector, the only way to ensure development and commercialization

TABLE 5-1   Patents Issued from 1979 to 1984

Patent Recipient	1979	1980	1981	1982	1983	1984
U.S.-based inventors						
(other than U.S. government)	33,391	36,978	42,050	38,092	34,129	40,857
U.S. government	992	1,156	1,144	1,007	993	1,205
Foreign-based inventors	21,035	23,093	27,816	26,053	24,593	30,087
Total U.S. patents	55,148	61,227	71,010	65,152	59,715	72,149

SOURCE: U.S. Commissioner of Patents and Trademarks, 1985. Annual Report Fiscal Year '84. U.S. Department of Commerce, Patent and Trademark Office. Washington, D.C.

of its discoveries may be through patents and exclusive licensing to the private sector. This arrangement might seem as though the licensee has received preferential access and control over the benefits of research supported by tax dollars. Nevertheless, the risk of unequal benefits must be weighed against the certainty that no benefit will be derived if a promising technology remains undeveloped.

Patents and the Federal Government

In 1980 approximately $62.7 billion was spent in the United States on R&D. The public and private sectors each contributed roughly half of this figure (NSF, 1983). Yet of the almost 70,000 patents issued annually in this country, the vast majority (97 percent) are awarded to the private sector (Table 5-1). Part of this disparity stems from the government's greater emphasis on basic research (18.7 percent of R&D expenditures vs. 4.1 percent for industry). However, another factor is government policy. Between 1973 and 1983, almost three-quarters of government patents were granted to only four agencies: the Air Force, Army, the National Aeronautics and Space Administration, and the Navy (Table 5-2)—not to transfer technology but to protect government procurement of goods produced under these patents (U.S. Commissioner of Patents and Trademarks, 1985).

Congress has emphasized patenting and licensing at other agencies through recent and pending legislation. The most significant acts were the Stevenson-Wydler Technology Innovation Act of 1980 (P.L. 96-480) and the Federal Technology Transfer Act of 1986 (P.L. 99-502), which mandated that technology transfer should be part of the missions of federal agencies and created mechanisms by which these agencies and their laboratories can transfer

TABLE 5-2   U.S. Government Agency Patents[a]

Agency	1980	1981	1982	1983	1984
USDA	54	53	46	45	46
Air Force	159	133	89	120	168
Army	233	229	196	205	200
Commerce	6	5	7	5	7
DOE	59	234	210	170	263
DOT	3	5	1	0	—
NSA	1	1	2	1	6
EPA	3	10	1	3	3
HHS	23	27	19	26	38
Interior	35	43	27	23	16
NASA	74	70	73	114	143
Navy	390	326	319	278	306
Postal Service	0	2	0	0	—
TVA	0	0	0	0	4
Treasury	0	2	1	1	1
VA	2	0	2	0	1
USA[b]	14	12	12	2	2
FCC	0	2	2	0	1
Total	1,156	1,144	1,007	993	1,205

NOTE: DOT = Dept. of Transportation; FCC = Federal Communications Commission; NSA = National Security Agency; TVA = Tennessee Valley Authority; and VA = Veterans Administration.

[a] These data represent utility patents assigned to agencies at the time of issue.
[b] No agency indicated.

SOURCE: U.S. Commissioner of Patents and Trademarks, 1985. Annual Report Fiscal Year '84. U.S. Department of Commerce, Patent and Trademark Office. Washington, D.C.

technology. Now inventions resulting from federally funded research with a cooperating private institution may, in general, be patented and an exclusive license granted by that institution. Recent levels of patenting by government agencies are shown in Table 5-2.

THE NATIONAL TECHNICAL INFORMATION SERVICE

The Department of Commerce's National Technical Information Service (NTIS) has played the leading role in marketing federally owned patents. The NTIS program covers some of the inventions created by the Departments of Commerce, Health and Human Services, Interior, Transportation, the Army and Air Force, and USDA, as well as those of the Veterans Administration and the Environmental Protection Agency. Recent federal legislation cited previously probably will relieve NTIS of some of this responsibility.

NTIS publicizes government inventions available for licensing, files for foreign patents, and negotiates licensing agreements that may involve exclusivity—that is, one licensee (or sometimes several) with exclusive use of a patent. Nonexclusive licenses are granted in cases in which access to a technology by many competing firms would not discourage commercialization. Table 5-3 shows the levels of licensing by NTIS since FY82 and projects them to 1990.

As part of the licensing negotiations, NTIS requires companies to file development plans for inventions. These plans specify the amount the licensee will invest in R&D, in seeking approval from regulators, and in commercialization. The pledge of capital investment ensures that the licensee is serious about developing the invention and is not buying the license simply to prevent competition with its own products. For the 77 licenses granted by NTIS in FY83 and FY84, licensees pledged a total of $178 million.

As noted earlier, NTIS has managed patents in agriculture and the biomedical sciences. Since the liberalization of federal patent policy under the Stevenson-Wydler Technology Innovation Act of 1980, the number of invention reports filed by NIH-funded universities has doubled. The Federal Technology Transfer Act of 1986 is expected to further increase the number of patents filed by federal employees. For example, NIH's scientists, working intramurally, now file about 150 invention reports per year. From these, NIH files about 50 patent applications per year. NTIS markets and manages NIH patents, and about 30 percent of NIH patents are eventually licensed. Although granting exclusive licenses on government-held patents might appear to stifle competition, licenses on certain drugs or other socially beneficial products may serve the public interest by encouraging private investment in research, development, and marketing.

ARS also uses NTIS to promote its patents. Of the approximately 50 patents filed per year by ARS scientists (Table 5-4), about half go to NTIS for licensing. USDA also promotes and licenses its own inventions. USDA requires licensees to specify the amount they will spend on commercialization. In 1985, $30 million was pledged to develop 30 ARS inventions.

TABLE 5-3 National Technical Information Service (NTIS) Patent Licensing Activities

Fiscal Year	Revenues (thousands of dollars)				Licenses	
	Execution Fees	Minimum Payment Fees	Running Use Fees	Total	Granted	In Force (end of year)
1982	45	41	69	155	27	76[a]
1983	78	59	770	907	41	117
1984	91	88	689	868	36	154
1985[b]	90	170	1,140	1,400	40	163
1986[b]	95	210	2,000	2,300	40	175
1987[b]	105	260	2,500	2,850	45	190
1988[b]	115	310	2,500	2,950	50	206
1989[b]	125	375	3,000	3,500	50	220
1990[b]	135	450	3,500	4,000	50	232

[a] Licenses issued prior to the establishment of the current NTIS program.
[b] Values for these years are estimates.

SOURCE: National Technical Information Service (1985).

TABLE 5-4   USDA Patent License Activities[a]

Activity	1979	1980	1981	1982	1983	1984	1985
Patents issued	39	53	55	45	45	46	39
Public inquiries	77	119	185	293	241	407	666
Nonexclusive licenses awarded	68	69	30	21	40	26	16
Exclusive licenses awarded	0	0	3	5	6	14	17
Annual reports received							
(nonexclusive licenses)	101	158	186	140	122	162	62
Patents transferred to Dept. of Commerce							
for exclusive negotiations	2	4	8	9	22	22	17

[a] Combined USDA–Agricultural Research Service activity.

SOURCE: Coordinator, National Patent Program, USDA, 1985.

Patents and Universities

The following paragraphs describe how two universities dealt with patenting by establishing their own formal programs. Variations on the first approach have been used at other universities, for example, the Purdue University Research Foundation, the Iowa State University Research Foundation, and the Research Corporation, which handles patenting and licensing for a number of universities.

WISCONSIN ALUMNI RESEARCH FOUNDATION

In 1925, nine alumni of the University of Wisconsin formed the Wisconsin Alumni Research Foundation (WARF). WARF was and is free of university control. It exists solely to support research and promote the discoveries of university faculty and students by underwriting the patenting and licensing process for these inventors.

The university itself holds no patents. Faculty members can choose between negotiating patents and licenses with commercial contributors themselves or giving that responsibility to WARF. Most choose the latter. After more than 50 years with this arrangement, the university has yet to report a conflict of interest.

Faculty inventors receive 15 percent of the royalties after costs on patents licensed by WARF; the remainder goes to the University of Wisconsin graduate school to support research projects. Although the university will not involve itself directly in patenting, it will withhold publishing research results for up to 90 days to facilitate filing a patent application. However, the university does not permit an indefinite delay of publication and insists on the freedom to communicate results—a tenable position, for only individual faculty members or WARF, not the university, can hold patents.

Two major patents—in terms of income from royalties—have emerged from the WARF program: a process for irradiating milk in order to activate vitamin D ($8 million net) and the discovery that led to the commercialization of coumarin (warfarin), an anticoagulant and rodenticide ($4 million net). In all, 42 income-producing inventions were assigned to WARF between 1925 and 1975, of which 12 earned more than $100,000 in net royalties. Since 1928, WARF has distributed $100 million earned from royalties and investments to the University of Wisconsin (Omenn, 1982b).

COLUMBIA UNIVERSITY SCIENCE AND TECHNOLOGY DEVELOPMENT OFFICE

As late as 1981, Columbia University had no policy on patenting. As a result, many technologies developed at the university were never exploited. The faculty was in general not entrepreneurial, and those who did negotiate deals with private industry tended to do so independently. This situation created a subculture of individual arrangements at Columbia that often put restrictions on research but offered little or no protection of intellectual property.

To combat these problems the university opened the Science and Technology Development Office in 1982. Its goals are to obtain patents on university inventions, license those inventions, and create a structure for interaction with the private sector that will feed money back into the university.

The Science and Technology Development Office has a policy committee that handles conflict of interest questions and an administrative committee that examines research proposals from a business standpoint. All proposals are initiated by Columbia researchers, and the funding company usually has rights to an exclusive license if a commercial product should result. There can be no delays imposed on publication—the company has 30–60 days to review early drafts. However, Columbia reserves the right to patent anything, regardless of the funder's recommendations.

This policy relieves the university from pressure to withhold information (at seminars, for example); however, such a policy also means the university may rush the patent process and obtain a patent that may ultimately be indefensible.

The Science and Technology Development Office has a yearly budget of $540,000. Of this amount, $123,000 goes to legal fees for filing patents. Companies that receive licenses on patents must also grant the university the right to approve sublicensing to other companies. The office is not directly interested in product development, however.

As of March 1985, the Science and Technology Development Office had generated $2 million—through investments, not royalties—which is channeled back into the university to support research. Although this amount is relatively small, it is the portion of Columbia's interactions with the private sector that is unburdened by restrictions attached to other kinds of private grants and gifts. Ideally, the office would have control of all private grants to the university.

Revenues from Licenses

Reliable data are not available on the license value of patents. However, it is generally accepted that the average royalty earnings of patents is low. A sample of patents awarded to 33 technology-oriented firms showed that 20 percent of the licenses earned less than $1,000 per year, 40 percent less than $5,000, 60 percent less than $10,000, and 95 percent less than $100,000 (Roberts, 1982).

The situation is similar in the public sector. For the 154 NTIS licenses in effect at the end of 1984, the average annual revenue was $5,636. As Table 5-3 shows, government revenues from the NTIS program are expected to grow from $868,000 in FY84 to $4 million in FY90. (This estimate may prove low, given the Federal Technology Transfer Act of 1986.) Revenues from licenses are returned to the U.S. Treasury, with a percentage going to the inventor. Recently, $40,000 was distributed to 100 inventors. Maximum payments were $8,000.

However, two biotechnology patents held by Stanford University and the University of California have already generated revenues in excess of $5 million for these institutions. The patents, issued in 1980, cover a process for making "biologically functional molecular chimaeras" (recombinant DNA) and products derived by this process. Currently 81 companies each pay $10,000 annually to license both the process and product patents. The universities also earn royalties ranging from 0.5 to 10.0 percent on commercial product sales, depending on the type of product being marketed. Patent revenues, divided equally among the inventors (S. Cohen and H. Boyer), their departments, and the schools, are used mainly for research and education at the universities. This example, outstanding in terms of its financial success, indicates the payoff potential of biotechnology patents.

Biotechnology Patenting Activity

Approximately 2 percent of recently granted U.S. patents cover biotechnology inventions (Table 5-5; OMEC International, 1985). Between 40 and 45 percent of these patents are granted to foreign individuals or organizations, roughly the same percentage as with all patents. About 40 percent of biotechnology patents are granted to U.S. corporations, and about 18 percent go to U.S. universities, government, nonprofits, and individuals.

Table 5-6 shows the levels of patenting activity for the 11 U.S. universities that accounted for most biotechnology inventions. Although biotechnology patents account for about 2 percent of all patents granted by the United States, for these universities they vary from 14 percent for Iowa State University to 37 percent for

TABLE 5-5   U.S. Biotechnology Patent Activity (Patents Issued)[a]

Activity	1983	1984
All patents[b]	59,715	72,149
U.S. corporate biotechnology	400	441
U.S. university biotechnology	68	95
Other U.S. (government, nonprofits, and individuals)	94	127
Total U.S.-based	562	663
Foreign corporate biotechnology	383	371
Other foreign biotechnology	73	80
Total foreign	456	451
Total biotechnology	1,018	1,114

[a]SOURCE: OMEC International, 1985. *Biotechnology Patent Digest* 4(10):150–151, unless otherwise indicated.

[b]SOURCE: U.S. Commissioner of Patents and Trademarks, 1985. Annual Report Fiscal Year '84. U.S. Department of Commerce, Patent and Trademark Office. Washington, D.C.

TABLE 5-6 Number of Biotechnology Patents Granted
to Selected U.S. Universities

Patent Recipient	Biotechnology Patents[a]		All Patents[b]
	1983	1984	1984
University of California	16	16	45
Massachusetts Institute of Technology	8	6	47
University of Wisconsin (WARF)[c]	3	6	16
Stanford University	2	6	16
Harvard University	8	5	NA
Cornell University	2	5	12[d]
Purdue University (Research Foundation)	1	4	14
University of Illinois	1	2	NA
Iowa State University (Research Foundation)	1	2	14
Montana State University	1	2	NA
Northwestern University	1	2	NA
All other	24	39	—
Total	68	95	—

NOTE: NA = not available.

[a] OMEC International, 1985. *Biotechnology Patent Digest* 4(10):150–151.

[b] *IPO News* 15(4):3, 1985.

[c] Wisconsin Agricultural Research Foundation.

[d] Cornell University Patent and Licensing Office, personal communication, 1985.

the University of Wisconsin (WARF). Thus, biotechnology patents have become a significant part of patenting activity at universities.

Patenting activity in biotechnology by private firms is an evolving field, still subject to considerable uncertainty. Publicly held biotechnology firms frequently address patent issues in their annual financial reports to stockholders and to the U.S. Securities and Exchange Commission (Form 10-K). Although biotechnology firms have different approaches to protecting their intellectual property, statements in these reports indicate that these firms seek patent protection only if they believe the patents will be valid and enforceable. If this does not seem likely, they try to keep such technology as trade secrets.

Nonpatented Intellectual Property

Basic research at universities spawns many innovations that cannot be patented but are valuable intellectual property and important components of technology transfer. The most amorphous components can be termed "know-how" and "show-how," intellectual advances and new techniques for research generated in university laboratories at the cutting edge of a scientific field. Industry expects this contribution from universities, just as it expects universities to train researchers to fill industry's laboratories. Much of industry's impetus to form university–industry partnerships, pay university faculty as consultants, and hire prominent scientists into industrial laboratories comes from its desire to gain access to "know-how" and "show-how" on new technology. These university contributions must therefore be recognized under the rubric of technology transfer.

Other forms of nonpatentable intellectual property are more tangible and can be licensed or copyrighted. Computer software developed by the public or private sector can be copyrighted. Important products of biotechnology research that can be licensed include specialized cell lines derived from animals, plants, or microbes that are used for basic research or product development.

Conclusions

Patenting and licensing play a necessary, if limited, role in advancing technology transfer from the public to the private sector. Exclusive licensing of government-funded inventions to industry is particularly important in areas such as biotechnology, because their commercialization potential will attract the private sector only if the reward for capital-intensive development is the sole right to manufacture and sell the product.

In addition, there is evidence that publicly owned patents serve as "technology building blocks." In a sample of food-related patents held by the USDA and private parties, USDA patents were cited proportionately more often in subsequent patent filings. Thus, even though federally owned patents may not always be directly commercialized, they may still contribute to future innovation (Evenson and Wright, 1980). The Federal Technology Transfer Act of 1986 should stimulate patenting and licensing by federal laboratories.

Limitations of patenting and licensing must not be forgotten, however. Few inventions produce major commercial wins; hence, licensing fees in both the public and private sectors produce modest returns. Furthermore, the delay between the award of a license and the actual practice of a patent can be as long as 10 years, and even though companies pledge funds for development, there is no guarantee of an eventual product. It is therefore more realistic to view the securing of patents and the assigning of licenses by the public sector as one of several instruments of technology transfer. Royalties from university or government patenting and licensing cannot be considered significant sources of revenue for reinvestment in basic research. However, public sector patenting has value in spurring innovative research directed toward practical ends, in promoting technology transfer from the public to private sector, and in providing supplemental income to research institutions. Currently, universities and government do not always fully exploit their patents because of poor incentives due to policies on distributing royalties. Industry's patent experience might offer the public sector a better model.

Public policy issues pertinent to biotechnology patents center around two main issues: uncertainty about the scope of protection provided by patents and the government's role in generating research results. There have been charges that excessively broad patents have been issued (Webber, 1984). If this is true, firms may be induced into socially undesirable patterns of R&D expenditure, and prolonged litigation and delays in commercialization can be expected.

Government and university research appear to lead to biotechnology patents in greater proportion to its investment than in other areas of science and technology. This is consistent with the focus on basic research by government and university laboratories and the basic research requirements of biotechnology. Besides raising the usual concerns over conflict of interest and freedom of research, this concentration of patenting activity focuses attention on organized mechanisms for transfer of technology to promote research, development, and their ultimate benefits for society.

RECOMMENDATIONS

Roles for Universities and Government Agencies

Universities and state and federal agencies are expanding both the nature and number of their relationships with the private sector as they explore ways to increase scientific communication and the flow of technology. The federal government, granting agencies, and public and private universities should encourage interdisciplinary research, partnerships, and new funding arrangements among universities, government, and industry. The Federal Technology Transfer Act of 1986 provides new incentives to federal scientists in this regard. Consultancies, affiliate programs, grants, consortia, research parks, and other forms of partnership between the public and private sectors that foster communication and technology transfer should be promoted. The USDA, SAESs, and CES should emulate other agencies such as NIH and NBS in forming innovative affiliations to increase technology transfer.

Cooperative Extension Service

The CES should focus some of its efforts on the transfer of biotechnology research that will prove adaptable and profitable to the agricultural community. It should train many of its specialists in biotechnology and increase its interactions with the private sector to keep abreast of new biotechnology valuable to the agricultural community. Furthermore, CES should work to anticipate and alleviate social and economic impacts that may result from the application of new biotechnologies. CES should also play a key role in educating the public about biotechnology.

Patenting and Licensing

Patenting and licensing play necessary roles in advancing technology transfer and assuring the commercialization of research results, especially in capital-intensive fields such as biotechnology. Patenting and licensing by universities and government agencies should be encouraged as one of several instruments used to transfer technology. Universities and government agencies should provide incentives to their scientists to encourage patenting. Public policy should encourage state land-grant universities to confer exclusive license on patents to private companies with the resources, marketing, and product interests required to translate these discoveries into commercial products.

Regulation of Environmental Testing

The government's uncertainty over appropriate regulatory steps has fueled public controversy over the assessment of possible environmental risks from genetically engineered agricultural products. The Food and Drug Administration, USDA, and EPA must formulate, publish, and implement a research and regulatory program that is based on sound scientific principles. Initially, 5–10 selected, already-existing publicly owned field stations should be available as an option for environmental release testing, professionally managed by an oversight committee of public sector scientists with expertise in agronomy, ecology, plant pathology, entomology, microbiology, molecular biosciences, and public health. This interim program should be designed to gain scientific information and practical experience with field testing and to protect the public safety. The current lack of adequate regulatory procedures is halting progress in applying biotechnologies to agriculture.

SUMMARY

America has traditionally been at the forefront of world agriculture. Our capacity to develop and implement new technology, as well as the bounty of our land and natural resources, are responsible for this. In a modern, changing world these facets—resources, expertise, technology, and application—remain of paramount importance.

Biotechnology offers us exciting new avenues to increase agricultural productivity. Its tools, combined with advances in the science of agricultural systems, can lead to more nutritious food produced more efficiently. We need this science and technology to maintain our competitiveness and world leadership.

The strategies for national competitiveness involve many players. We must increase the emphasis on basic research in our schools of agriculture and public and private universities. We must improve the techniques and applications of science. We must promote these goals by integrating research across disciplines and institutions and by assessing projects through peer and merit review. We must train enough research personnel and extension agents to conduct research and applications of biotechnology in agriculture. We must encourage technology transfer through government–university–industry relationships and patenting activities. And we must formulate workable guidelines and procedures for environmental testing of biotechnology products. Our federal and state governments, public and private universities, and private sector institutions and industries all have important roles to play in achieving these goals for agriculture.

References

Agricultural Research Institute. 1985. A Survey of U.S. Agricultural Research by Private Industry III. Bethesda, Md.: Agricultural Research Institute.

Anderson, C. J. 1984. Plant Biology Personnel and Training at Doctorate-Granting Institutions. Higher Education Panel Reports No. 62. Washington, D.C.: American Council on Education.

Blumenthal, D., M. Gluck, K. S. Louis, and D. Wise. 1986. Industrial support of university research in biotechnology. Science 231:242–246.

Bonnen, J. J. 1983. Historical sources of U.S. agricultural productivity: implications for R&D policy and social science research. Am. J. Agric. Eco. 65:958–966.

Brown, A. W. A., T. C. Byerly, M. Gibbs, and A. San Pietro. 1975. Crop Productivity—Research Imperatives. Paper presented at an international conference held at Boyne Highlands Inn, Harbor Springs, Mich., October 20–24, 1975.

Buttel, F. H., M. Kenney, and J. Kloppenburg. 1985. Industry–University Relationships and the Land-Grant System. Paper presented at a Symposium on the Agricultural Scientific Enterprise, University of Kentucky, Lexington, March 1985.

Carson, R. 1962. Silent Spring. Boston, Mass.: Houghton Mifflin.

Cole, S., L. Rubin, and J. R. Cole. 1978. Peer Review in the National Science Foundation, National Research Council. Washington, D.C.: National Academy Press.

Evenson, R. E., P. E. Waggoner, and V. W. Ruttan. 1979. Economic benefits from research: an example from agriculture. Science 205:1101–1107.

Evenson, R. E., and B. Wright. 1980. An Evaluation of Methods for Examining the Quality of Agricultural Research. U.S. Food and Agricultural Research Paper No. 6. Washington, D.C.: Office of Technology Assessment.

145

Experiment Station Committee on Organization and Policy. 1984. Research 1984. The State Agricultural Experiment Stations. Cooperative State Research Service. Washington, D.C.: U.S. Department of Agriculture.

Government–University–Industry Research Roundtable. 1986. New Alliances and Partnerships in American Science and Engineering. Washington, D.C.: National Academy Press.

Hightower, J. 1973. Hard Tomatoes, Hard Times. Cambridge, Mass.: Schenkman.

Institute of Medicine. 1985. Personnel Needs and Training for Biomedical and Behavioral Research. Washington, D.C.: National Academy Press.

Institute of Medicine. 1983. Personnel Needs and Training for Biomedical and Behavioral Research. Washington, D.C.: National Academy Press.

Intersociety Working Group. 1986. AAAS Report XI. Research and Development, FY 1987. Washington, D.C.: American Association for the Advancement of Science.

Kiesler, S. 1986. Thinking ahead, the hidden messages in computer networks. Harvard Bus. Rev. 64:46–59.

Magrath, W. B. 1985. Factors Affecting the Location of the U.S. Biotechnology Industry. Agricultural Economics Staff Paper 85–26, Cornell University, Ithaca, N.Y.

Mayer, A., and J. Mayer. 1974. Agriculture, the island empire. Daedalus 103:83–95.

National Agricultural Research and Extension Users Advisory Board. 1986. Appraisal of the Proposed 1987 Budget for Food and Agricultural Sciences. Report to the President and Congress. Washington, D.C.: U.S. Department of Agriculture.

National Association of State Universities and Land-Grant Colleges. 1986. Emerging Biotechnologies in Agriculture: Issues and Policies. Progress Report V. Washington, D.C.: NASULGC.

National Association of State Universities and Land-Grant Colleges. 1985. Emerging Biotechnologies in Agriculture: Issues and Policies. Progress Report IV. Washington, D.C.: NASULGC.

National Association of State Universities and Land-Grant Colleges. 1984. Emerging Biotechnologies in Agriculture: Issues and Policies. Progress Report III. Washington, D.C.: NASULGC.

National Association of State Universities and Land-Grant Colleges. 1983. Emerging Biotechnologies in Agriculture: Issues and Policies. Progress Report II. Washington, D.C.: NASULGC.

National Center for Education Statistics. 1985. The Condition of Education. Washington, D.C.: U.S. Department of Education.

National Research Council. 1985a. New Directions for Biosciences Research in Agriculture: High-Reward Opportunities. Washington, D.C.: National Academy Press.

National Research Council. 1985b. Report of the Research Briefing Panel on Agricultural Research. In New Pathways in Science and Technology: Collected Research Briefings 1982–1984, COSEPUP (pp. 62–87). New York: Random House.

National Research Council. 1984. Genetic Engineering of Plants: Agricultural Research Opportunities and Policy Concerns. Proceedings of a convocation sponsored by the Board of Agriculture, National Research Council. Washington, D.C.: National Academy Press.

National Research Council. 1983. Summary Report 1983, Doctorate Recipients from United States Universities. Washington, D.C.: National Academy Press.

National Research Council. 1975. World Food and Nutrition Study: Enhancement of Food Production for the United States. Washington, D.C.: National Academy Press.

National Research Council. 1972. Report of the Committee on Research Advisory to the U.S. Department of Agriculture. Washington, D.C.: National Academy Press.

National Science Board. 1985. Science Indicators: The 1985 Report. Washington, D.C.: National Science Foundation.

National Science Board Commission on Precollege Education in Mathematics, Science and Technology. 1983. Educating Americans for the 21st Century. Washington, D.C.: National Science Foundation.

National Science Foundation. 1986. Federal Funding for Research and Development: Federal Obligations for Research to Universities and Colleges FY 1973–1986. Washington, D.C.: National Science Foundation.

National Science Foundation. 1983. Science Indicators 1982. Washington, D.C.: U.S. Government Printing Office.

Office of Science and Technology Policy. 1986. Coordinated framework for regulation of biotechnology: announcement of policy and notice for public comment. Fed. Regist. 51(123):23301–23393.

Office of Technology Assessment. 1986. Technology, Public Policy, and the Changing Structure of American Agriculture (OTA-F-285). Washington, D.C.: U.S. Government Printing Office.

Office of Technology Assessment. 1984. Commercial Biotechnology: An International Analysis. Washington, D.C.: U.S. Government Printing Office.

Office of Technology Assessment. 1983. Commercial Biotechnology. An International Analysis (0-25-561). Washington, D.C.: U.S. Government Printing Office.

Office of Technology Assessment. 1981. An Assessment of the U.S. Food and Agricultural Research System (OTA-F-155). Washington, D.C.: U.S. Government Printing Office.

OMEC International. 1985. Patent analysis. Biotechnol. Patent Dig. 4(10):150–151.

Omenn, G. S. 1982a. University–Corporate Relations in Science and Technology. Paper presented at the National Conference on University–Corporate Relations in Science and Technology, University of Pennsylvania, Philadelphia, 15 December 1982.

Omenn, G. S. 1982b. Taking university research into the marketplace. New Engl. J. Med. 307:694–700.

Press, F. 1986. Science: the best and worst of times. Science 231:1351–1352.

Roberts, E. B. 1982. Is licensing an effective alternative? Res. Manage. 9:20–24.

Ruttan, V. W. 1982. Agricultural Research Policy. Minneapolis: University of Minnesota Press.

Sitting, M., and R. Noyes. 1985. Genetic Engineering and Biotechnology Firms Worldwide Directory. Kingston, N.J.: Sitting and Noyes.

U.S. Commissioner of Patents and Trademarks. 1985. Annual Report Fiscal Year '84. Washington, D.C.: U.S. Department of Commerce, Patent and Trademark Office.

U.S. Department of Agriculture. 1986. Inventory of Agricultural Research Fiscal Year 1985. Washington, D.C.: U.S. Department of Agriculture.

U.S. General Accounting Office. 1986. Biotechnology: Analysis of Federally Funded Research (GAO/RCED-86-187). Washington, D.C.: U.S. General Accounting Office.

U.S. General Accounting Office. 1985. Biotechnology: The U.S. Department of Agriculture's Biotechnology Research Efforts (GAO/RCED-86-39-BR). Washington, D.C.: U.S. General Accounting Office.

Webber, D. 1984. Biotechnology firms gird for clash over patent claims. Chem. Eng. News 12:18–24.

White House Science Council Panel on the Health of U.S. Colleges and Universities. 1986. A Renewed Partnership. Office of Science and Technology Policy. Washington, D.C.: U.S. Government Printing Office.

Winrock International. 1982. Science for Agriculture: Report of a Workshop on Critical Issues in American Agricultural Research. New York: Rockefeller Foundation.

APPENDIX
Gene Transfer Methods Applicable to Agricultural Organisms

Phyllis B. Moses

INTRODUCTION

The transfer of genes from one organism to another is a natural process that creates variation in biological traits. This fact underlies all attempts to improve agriculturally important species, whether through traditional agricultural breeding or through the techniques of molecular biology. In both cases, human beings manipulate a naturally occurring process to produce varieties of organisms that display desired traits, for example, food animals with a higher proportion of muscle to fat, or disease-resistant corn.

The major differences between traditional agricultural breeding and molecular biological methods of gene transfer lie neither in aims nor in processes, but rather in speed, precision, reliability, and scope. When traditional, or classical, breeders cross two sexually reproducing plants or animals, they mix tens of thousands of genes in the hope of obtaining progeny with the desired trait or traits. Through the fusion of sperm and egg, each parent contributes half of its genome (an organism's entire repertoire of genes) to its offspring, but the composition of that half varies in each parental sex cell and hence in each cross. In addition, because the traits desired usually come from only one parent and may be controlled by one or a few genes, many crosses are necessary before the "right" chance recombination of genes results in expression of the trait in the offspring. Even then, the progeny usually have to be crossed back to the parental variety to ensure stable adoption of

the new trait. Sometimes undesired traits derived from one parent
of a new, improved variety persist whereas the desired traits are
lost.

Such are the difficulties and limitations of classical breeding.
Molecular biological methods of gene transfer alleviate some of
these problems by allowing the process to be manipulated at a
more fundamental level. Instead of gambling on recombination of
large numbers of genes, scientists can insert individual genes for
specific traits directly into an established genome. They can also
control the way in which these genes express themselves in the
new variety of plant or animal. In short, by homing in on desired
traits, molecular gene transfer can shorten the breeding time for
new varieties and, in addition, lead to improvements not possible
by traditional breeding.

Laboratory methods to move individual genes between organisms capitalize on naturally occurring mechanisms of gene transfer
other than sexual reproduction. These include uptake of DNA by
cells and cell-to-cell transfer of packaged genetic material such as
viruses. Scientists began by studying these mechanisms in simple systems—bacteria and the viruses that infect them. Research
has progressed at a remarkable rate. Now scientists can transfer genes into organisms as diverse as soybeans and sheep. Much
work remains, however, to perfect gene transfer and its attendant
technologies of embryo culture and plant regeneration.

Scientists have relied heavily on favorite model organisms such
as the bacterium *Echerischia coli* and the fruit fly *Drosophila
melanogaster*, because of their ease of manipulation and the large
body of scientific knowledge accumulated about them. Model
systems are critical to the progress of research. Nevertheless,
molecular biologists must extend their techniques to commercially
important agricultural organisms. Movement in this direction will
not replace all traditional agricultural breeding with molecular
gene transfer. It will, however, expand the array of methods
available to improve agriculturally important species.

General Considerations

Gene transfer occurs naturally among bacteria by a variety of
mechanisms. Scientists learned in the 1950s and 1960s to exploit
these mechanisms to study gene regulation in bacteria and in

the 1970s developed additional artificial gene transfer methods for bacteria. It turned out to be a relatively simple matter to get some types of bacteria to take up pieces of DNA from their surrounding medium. Genes contained on the new pieces of DNA could be stably inherited and expressed to give new characteristics to the host bacteria. Scientists then devised special conditions that improved DNA uptake, maintenance, and gene expression in the new hosts. Gene transfer is now a routine laboratory procedure for bacterial strains such as *E. coli*.

The goals of gene transfer experiments with other organisms are the same as those of earlier work with bacteria—to study gene regulation and to obtain stable inheritance and expression of new characteristics. The difference is that these other organisms are more complicated biological entities than are bacteria. Hence the experimental problems and procedures are more complicated. It has proved necessary to devise some very special conditions and tools to move DNA into the cells of other organisms.

The explosion of knowledge in molecular biology is the direct result of certain basic biological discoveries that permit scientists to handle genes as macromolecules. Researchers can identify, isolate, cut, and splice genes and transfer them from one species to another. Enzymes, obtained mainly from bacteria, enable scientists to perform the first four steps on genes from any organism, by procedures that are now standard in molecular biology. The fifth step, gene transfer, must be worked out individually for different organisms.

Different species of animals, plants, and microbes vary widely in the ease with which gene transfer can currently be carried out. Plants have in general been more difficult to deal with than animals or microbes. The technology is improving rapidly, however, and it is likely that most organisms will in time be tractable targets for gene transfer.

This report surveys the scientific status and short-term prognosis for gene transfer systems applicable to animals, plants, and microbes of agricultural importance. The development of these methods has hinged on scientists' understanding of underlying molecular mechanisms in organisms. These mechanisms are then exploited, as was originally done for bacteria, to develop methods

for moving genes between organisms. This permits both fundamental studies of gene function and the endowment of an organism
with new, desired characteristics.

Many factors must be considered in the design of gene transfer
systems. The first requirement is an easily detected "tag" for the
gene so that its progress into a new host can be traced. Sometimes
the uniqueness of the foreign gene is sufficient: The gene can be
identified by the new characteristic it confers or its physical presence can be detected by a probe for its particular DNA sequence.
Unfortunately, such direct identification methods are sometimes
either impracticable or inconvenient and time consuming. In these
cases the foreign gene can be tagged by attaching it to another
gene whose presence in the host is easily and rapidly detectable.
Genes for drug resistance are often used as tags. Host cells that
incorporate these genes—along with the foreign gene—can survive
a drug treatment, whereas cells that have not taken up the genes
will die.

Another important consideration is the efficiency of gene
transfer. The probability of success must be high enough for
transfer of the gene to be detected with a reasonable frequency. If
drug resistance or other selection schemes are used, a lower frequency may be acceptable. In such cases many cells can be treated
for gene transfer, but only those few that actually incorporate the
foreign genes survive the selective treatment and are recovered.

Special vectors can improve the efficiency of gene transfer.
Foreign genes attached to the vector will be carried by it into the
host cell. Vectors are often derived from circular DNA molecules
called plasmids, or from viruses.

Different transfer systems have particular features that can
limit the size of foreign DNA segments that they are capable of
transferring. Segment sizes are measured in base pairs, the fundamental chemical units of DNA. A typical gene may be composed
of anywhere from 1,000 to 50,000 base pairs. A few techniques for
gene transfer can handle segments of DNA at the upper end of this
range, but most current methods are limited to segments at the
lower end. Because it may be desirable to transfer more than one
gene at a time, scientists are working to develop more and better
vectors that can handle multiple genes.

When a DNA molecule is used as a vector for foreign genes, its
own size sometimes limits how much extra DNA it can carry. Small

vectors, themselves less than 10,000 base pairs, are more often limited than are large vectors, which may be well over 100,000 base pairs. However, larger vectors require extra manipulations to equip them with foreign DNA.

The final state of foreign genes inside the host cell is also important. Genes can be maintained on vectors that are independent, self-replicating "minichromosomes," or they can be integrated into the larger chromosomes of the host cell. Depending on the experiment's purpose, independent maintenance or integration of new genes may be preferable. However, to ensure stable inheritance of transferred genes in intact animals or plants, the genes must usually be integrated.

Related to the state of genes is the question of gene copy number, that is, how many identical copies of a foreign gene end up in the cell. Again, depending on the experiment's purpose, many copies or only one copy may be desired. For example, if gene transfer is used to engineer a cell line to manufacture large amounts of a commercially valuable protein, a high copy number, self-replicating minichromosome would be used. On the other hand, if the purpose is to equip an animal or plant with a new gene for disease resistance, only one or two copies of the gene might be needed in the organism's own chromosomes, where it would be properly expressed and inherited by succeeding generations.

Transferred genes must be regulated so that their protein products are made in appropriate amounts at the correct time and in the right place. Genes are normally controlled by certain sequences in the surrounding DNA. These sequences in turn are affected by various factors within cells, for example, hormones. Transferred genes can be regulated by their normal control sequences. Alternatively, scientists can equip them with new control sequences to either mimic the natural situation or achieve new effects.

Before permanent genetic modification of an organism is attempted, it is important to study the gene of interest under various conditions to understand its normal function and regulation and to engineer any beneficial changes. These studies are conveniently done using a "transient expression" system, by which the activity of transferred genes can be rapidly measured inside cells, without waiting for stable, long-term genetic modification of the cells.

The flexibility and rapidity of initial studies are also enhanced by "shuttle vectors," which are designed to replicate in both animals or plants and in bacteria. By using shuttle vectors, scientists can easily grow and isolate genes in quantity from bacteria, modify them in vitro, and then quickly transfer them into animal or plant cells to test their function. The transferred genes can also be reisolated from the animal or plant cells, put back into bacteria, and grown in quantity again for further use.

DIRECT DNA UPTAKE

The earliest and still most widely used method for introducing DNA into animal cells grown in culture in the laboratory is direct uptake of DNA from the surrounding culture medium. The conditions are in principle the same as those used for bacterial cells: DNA must enter the cell and become stably maintained and inherited in the cell line in such a way that its new genetic information is expressed to confer a new trait on the cell.

The mechanics differ because animal cells differ structurally from bacterial cells. On the one hand, animal cells have only a membrane surrounding their contents, whereas bacterial cells (and plant, fungal, and yeast cells) have both a membrane and a wall. The rigid cell wall of the latter organisms often must be removed to allow DNA to enter the cell. On the other hand, most of the genetic information in animal, plant, fungal, and yeast cells is sequestered in the nucleus, an organelle surrounded by its own membrane. (Organisms that have cell nuclei are known as eucaryotes.) New genetic material usually must pass through this second membrane in order to be permanently added to a eucaryotic cell. Bacteria (known as procaryotes) lack an organized nucleus and usually accept new DNA more easily.

The major advantages of direct DNA uptake (facilitated by chemical or electrical treatments, as will be described) are its simplicity and applicability to many organisms and cell types. Hundreds of thousands of cells may be simultaneously treated, in contrast to microinjection of DNA into individual cells (described later), which is laborious and time consuming. Because it is so simple and rapid, direct uptake is extremely useful for basic studies of gene expression in cell culture. These studies are important for characterizing a gene's function, before researchers attempt elaborate and time-consuming gene transfer experiments in whole animals or plants.

Foreign genes introduced by direct uptake are expressed in their new host cells after a short period, usually 1 or 2 days. Direct DNA uptake thus quickly reveals the function of newly isolated or engineered genes during this period of "transient expression." For long-term studies the genes must integrate into the cell's own chromosomes, or be carried in by the uptake of new chromosomes, to ensure that they are stably inherited. Integration occurs at a high frequency after direct DNA uptake into animal cells because so many copies of the foreign genes have been introduced. (Maintenance on new chromosomes is discussed in the sections on Cell Fusion and Vector-Mediated Gene Transfer.)

In addition, gene transfer into cultured cells by direct DNA uptake is used for the commercial production of genetically engineered proteins. Drugs, hormones, food additives, and other valuable substances can be manufactured by cells into which the appropriate genes have been transferred. Human insulin for treatment of diabetics is now manufactured in bacteria in this way.

The limitations of direct uptake, particularly for animals, center on the fact that intact organisms usually are not suitable recipients. Thus, gene transfer into an animal embryo usually must be accomplished by other means. For plants this is not a strict limitation, as many species can be regenerated into whole plants from a single cultured cell.

Chemical Treatments

Chemical treatments can induce animal cells to take up DNA from their medium; most frequently these cells are in culture rather than in living animals. In the simplest and most popular method, cells are mixed with DNA that has been precipitated with calcium phosphate (Graham and van der Eb, 1973). This treatment compacts the DNA, so cells take up many copies of the foreign genes. Alternatively, the chemical DEAE-dextran may be used to facilitate DNA uptake (McCutchan and Pagano, 1968).

Cells in culture are relatively unspecialized and often do not correctly regulate genes as would the specialized organs of an intact animal. Researchers have therefore developed a technique to introduce DNA directly into intact organs, such as the liver or spleen, of living animals. Calcium phosphate-precipitated DNA is injected directly into the organs, in combination with low concentrations of enzymes that allow the DNA to enter (Dubensky et al., 1984). This technique enables researchers to quickly study an isolated gene's function in the differentiated, specialized cells of an intact organ, which more accurately reflect the gene's proper function in an animal. A variation of the organ transformation technique involves injecting the calcium phosphate-precipitated DNA intraperitoneally, where it is taken up and expressed by animal tissues, such as those of the liver and spleen (Benvenisty and Reshef, 1986).

Plant cells have been difficult to transform by chemical methods, but recently breakthroughs have been made. Polyethylene glycol has been used to obtain direct uptake and stable maintenance of DNA by protoplasts from a species of wheat, *Triticum monococcum* (Lörz et al., 1985), another monocot grass, *Lolium multiforum* (Potrykus et al., 1985a), and the dicots oilseed rape, tobacco, and petunias (Potrykus et al., 1985b). The frequency of integration of DNA after direct uptake is sometimes lower than for vector-mediated gene transfer into plants (discussed later), but there are no species restrictions on the type of host cell. However, protoplasts are used as recipients, so they must be capable of regenerating into plants for direct uptake to yield genetically altered species for agriculture.

Insect, fungal, yeast, and bacterial cells are all amenable to variations of calcium phosphate or other chemical treatments for direct DNA uptake. Often, direct uptake is used to introduce vector DNA molecules containing engineered genes. Direct uptake procedures simply place foreign genes inside the cell; vectors can help to integrate the genes into the cell's chromosomes or stably maintain the genes within the cell on the vector's minichromosome (see the section on Vector-Mediated Gene Transfer).

Electroporation

A newer method that is being widely adopted is electroporation (Neumann et al., 1982; Potter et al., 1984). Cells are mixed with DNA in solution and subjected to a brief pulse of electrical current. It is thought that the current pulse creates transient pores in the cell's membrane that allow DNA to enter efficiently.

Electroporation may work for any type of cell, even those that have resisted DNA uptake by chemical treatments, for example, cells of the immune system.

Electroporation can introduce DNA into protoplasts of both major categories of plants—dicots (e.g., carrots and tobacco) and monocots (e.g., corn; Fromm et al., 1985). Electroporation provides a transient gene expression system for plants. As discussed previously, transient expression systems are very useful for preliminary characterization of new genes. The lack of such a system for plants had previously held up progress in characterizing plant genes. Electroporation also permits stable integration of genes into plant chromosomes. It has been used successfully to stably transform corn and tobacco cells (Fromm et al., 1986; Schocher et al., 1986; Shillito et al., 1985).

DNA MICROINJECTION

DNA can be injected directly into single living cells using very fine glass pipettes (hollow needles). Experimenters use an elaborate apparatus consisting of a microscope and delicate micromanipulators to view the cell, hold it steady, and inject a solution containing DNA. As with chemical or electrical uptake methods, foreign genes can be in the form of isolated molecules or attached to vectors. A disadvantage compared to direct uptake is that relatively few individual cells can be injected; however, the frequency of successful incorporation of DNA per injected cell is higher.

Animals

Microinjection has been very successful for delivering foreign genes into mouse embryos at an early stage of development. Usually DNA is injected directly into a particular structure, the male pronucleus, of a fertilized mouse egg. This is the most receptive structure to the incorporation of foreign DNA. The embryos are subsequently reimplanted into foster mothers for development to term. Foreign genes are incorporated into the developing cells' chromosomes and are often present in every cell of the mature animal. Animals given new genes by this procedure are called "transgenic." Their new genes are usually passed on normally to their progeny. These foreign genes can be expressed, that is, make their protein products, which can confer new characteristics on the animal. The now classic example is transgenic mice containing foreign genes for growth hormone. Expression of these genes caused the mice to grow to up to twice their normal size (Hammer et al., 1984; Palmiter et al., 1983).

Many other animal genes have now been transferred into fertilized mouse eggs by microinjection and correctly expressed in the resulting mature mice. These include the chicken transferrin gene expressed in the liver (McKnight et al., 1983); a mouse immunoglobulin gene expressed in the spleen (Brinster et al., 1983); the rat elastase gene expressed in the pancreas (Swift et al., 1984); the rat skeletal muscle myosin gene expressed in skeletal muscle (Shani, 1985); a chimaeric mouse/human β-globin gene in blood, bone marrow, and spleen (Chada et al., 1985); and a swine histocompatibility gene (Frels et al., 1985).

Traits of potential economic value to the farmer that might be transferred by microinjection include increased levels of certain circulating hormones, antibiotic resistance, and immunoglobulins (antibodies) for "genetic vaccination" against pathogens. As noted previously, the introduction and expression of such genes has been successful in mice.

A necessary supporting technology for in vitro microinjection of mammalian embryos is embryo transfer into surrogate mother animals, for in vivo development of the embryos to term. Embryos of each livestock species must be handled in a slightly different manner, which must be experimentally determined.

Hammer and his collegues (1985) reported the successful production of transgenic farm animals (rabbits, sheep, and pigs) by microinjection. The same foreign gene for growth hormone used to produce transgenic mice was used for these other species. New techniques were needed to visualize pronuclei for microinjection, because of differences in the fertilized eggs of each species. The microinjected gene was integrated into the chromosomes of all three species, and was expressed in some of the transgenic rabbits and pigs.

Scientists have been very successful in microinjecting genes into embryos of the laboratory fruit fly *Drosophila melanogaster* for studies on the molecular biology of this insect. Rubin and Spradling (1982) pioneered this approach with their transposable P-element vector (discussed in the section on Vector-Mediated Gene Transfer). This vector or others similar to it might be adapted for both beneficial and harmful insects of agronomic importance.

Researchers routinely microinject genes into frog eggs, which are very large and metabolically active cells, for basic studies on gene expression in animals. More recently, microinjection was used to transfer DNA into the chromosomes of developing fish eggs (Chourrout et al., 1986; Zhu et al., 1985). Projects are aimed at basic studies of fish molecular biology and questions of how fish respond to their environment at the molecular level, as well as at aquacultural applications.

Both bovine and fish growth hormone genes have been injected into fish eggs. It has already been shown that injection of the purified protein hormones augments fish growth (Gill et al., 1985; Sekine et al., 1985). Transferred genes should be even more effective than purified hormones in promoting fish growth. Researchers have injected metallothionein genes from both mammals and fish into fish eggs, with the goal of engineering fish resistant to toxic metals. They have injected "antifreeze" genes obtained from winter flounder (also found in all antarctic fishes) to increase the cold tolerance of commercially valuable fish.

Plants

Microinjection can be used to deliver genetic material into plant cells. Segments of DNA, whole chromosomes, and even cellular organelles such as chloroplasts, which contain their own DNA molecules, can be microinjected by methods used for animal cells, although certain physical properties of plant cells complicate the technique.

Key elements for protoplast microinjection include microscopic resolution of the cell nucleus, which is enhanced by staining with dyes; immobilization of the cell by a holding pipette, embedding within agarose, or adhesion to glass surfaces; and efficient cell culture techniques. Researchers can successfully transform up to 14 percent of the cells they microinject with DNA (Crossway et al., 1986). This high frequency might be increased further by using microinjection in conjunction with specially developed vectors, derived from the Ti plasmid or plant transposable elements (see sections on these vectors). Because of the high transformation frequency possible with microinjection, a direct selection scheme (e.g., drug resistance) is unnecessary. Furthermore, specific host-range requirements associated with the Ti plasmid or viral vectors are obviated.

Although at present the recipient plant species must be amenable to cell culture and regeneration from protoplasts, suspension cultures or pollen grains may be used in the future, which would bypass the problem of regeneration. Alternatively, DNA may be injected into the developing floral side-shoots of plants, where it can pass into germ cells. Researchers have reported that the cereal rye (a monocot) can be transformed in this way (de la Peña et al., 1987).

Microinjection of individual chromosomes or cellular organelles (e.g., chloroplasts, mitochondria, and nuclei) could potentially produce improved cultivars with new traits such as herbicide resistance or cytoplasmic male sterility. Transfer of traits by microinjection would be more direct, precise, and faster than by breeding or cell fusion (described in the next section), because microinjection transfers a specific, limited amount of genetic information. There would be less need for selection or backcrossing, which are often time-consuming, difficult processes.

Most agronomic traits are polygenic, that is, they are caused by the interplay of several different genes in the plant. Genetic studies often reveal that these genes are linked in blocks on specific segments of chromosomes. Classical plant breeding can sometimes transfer such traits between species via interspecific crosses, but these crosses are not always successful. Transfer of individual chromosomes would permit researchers to introduce traits that result from the interaction of several genes linked on that chromosome.

Chromosome microinjection would also enable the transfer of traits that are encoded by single genes that have not yet been identified and isolated. Much of the sophisticated biochemistry and genetics of single-gene traits known for animals and used to isolate important genes is lacking for plants. Consequently, few plant genes of agronomic importance have been isolated. Whole-chromosome transfer may allow scientists to genetically engineer plants that would not be tractable at this time by more sophisticated gene-splicing (recombinant DNA) techniques. Attempts are being made to transform plant cells by microinjection of isolated chromosomes (Greisbach, 1983, 1987).

CELL FUSION

Cell fusion combines the entire genetic contents of two cells, producing hybrid cells that often express certain traits from both parents. The parent cells can be from different species or from different types of the same species. Fusion is usually mediated by chemicals such as polyethylene glycol or dimethylsulfoxide, although newer techniques use electrofusion.

Animal Cells

Cell fusion is the basis for the manufacture of monoclonal antibodies. Monoclonal antibody-producing cell lines (hybridomas) are created by fusing antibody-producing B-cells from animals with myeloma cells, which grow indefinitely in culture. The pure, highly specific antibodies thus obtained are important reagents for research, medicine, and agriculture. Diagnostic kits and vaccines for animal health based on monoclonal antibodies are already on the market (Gamble, 1986). Diagnosis of plant pathogens such as viruses, bacteria, fungi, and nematodes can also be facilitated by tests based on monoclonal antibodies; commercial products should be available in the near future (Gamble, 1986).

Certain agricultural applications have been held back by lack of suitable myeloma lines for fusion with B-cells from farm animals, as opposed to standard laboratory animals such as the mouse. However, this problem can be surmounted by creating hybridomas by direct DNA uptake. DNA from B-cells and myeloma cells is simultaneously introduced into recipient cells by calcium phosphate coprecipitation or by electroporation (Gamble, 1986). This approach obviates the need to fuse interspecific cell lines, and thus solves the problem of finding suitable myeloma lines for different livestock species.

Fusion of animal cell lines in culture is also exploited to map genes to specific chromosomes, an important step in locating genes to use in transfer experiments and in breeding strategies. Gene maps for mice and men are quite advanced. Those for livestock lag behind, but efforts are starting, notably for swine (Fries and Ruddle, 1986). To map these genes, swine cells are fused to mouse cells in culture. The interspecies cell hybrids reject most of the swine chromosomes. Ideally, a set of cell lines, each harboring a single different swine chromosome, is made. Known DNA sequences are used as probes for particular genes with those sequences. These probes bind to defined lengths of DNA from the fused cells. Because swine and mouse chromosomes can be distinguished by small differences in DNA sequences (known as restriction fragment length polymorphisms), differences in the lengths of DNA containing the gene detected by the probe indicate whether that gene is on a swine or a mouse chromosome of the hybrid cell. Location on a swine chromosome pinpoints the gene to that single particular swine chromosome, which is the only swine chromosome in the hybrid cell. Gene mapping is expected to play an important role in finding genes for transfer of complex traits in livestock, such as lactation, fertility, growth, and disease resistance.

Plant Cells

In eucaryotic cells the cytoplasm—that part of the cell surrounding the nucleus—contains organelles that have their own separate DNA. In plants, protoplast fusion is used to transfer genes from both the nucleus and the cytoplasm. Fusion combines the genomes of two parents, as in traditional breeding, but results can sometimes be obtained faster, even though the fusion product must be backcrossed to the recipient line for several generations to create a new, stable line possessing the one trait desired from the donor. Protoplast fusion can be used for transferring genes that are hard to identify, isolate, and clone or for polygenic traits. Furthermore, protoplast fusion can be used for plants that cannot be crossed sexually (although plants regenerated from such fused hybrids may sometimes be sterile).

Most commonly, cells from closely related plants are fused in order to transfer one particular trait from the donor plant into the recipient. For example, a single dominant nuclear gene for resistance to tobacco mosaic virus (Evans et al., 1981) and a polygenic trait for hornworm resistance (Bravo and Evans, 1985) were transferred into tobacco lines by this method. Traits from a wild species can be introduced into a related cultivated species. Cells of wild and cultivated potato plants were fused to transfer the wild species' resistance to potato leaf roll virus (Austin et al., 1985). The hybrids were fertile, bore tubers like those of the cultivated species, and were resistant to the virus.

Cytoplasmic (mitochondrial and chloroplast) traits can be transferred by fusing a donor cell whose nucleus has been inactivated, usually by irradiation, with an intact recipient cell to form a "cybrid." Initially, the cybrid contains the active nucleus of the recipient cell along with mitochondria and chloroplasts from both the donor and recipient cells. However, progeny cells that contain mitochondrial or chloroplast genotypes from one parent only quickly segregate. Plants are then regenerated from cells that harbor the desired donor cytoplasmic genotypes. Both cytoplasmic male sterility (mitochondria) and resistance to the triazine class of herbicides (chloroplast) have been transferred into a single *Brassica* line via cybrid formation (Pelletier et al., 1983).

VECTOR-MEDIATED GENE TRANSFER

A vector is a molecule of DNA that is attached to a foreign gene to facilitate its transfer, maintenance, and expression within the target cell. Vectors offer many advantages: high frequency of gene transfer, transfer into specific cell types, more control over the final copy number of a transferred gene, and certain properties that make them easy to track, permit them to be stably maintained in the target cell, and enable them to express foreign genes. Vectors can, therefore, greatly improve gene transfer. However, different species and cell types may require different types of vectors, and often much work must go into creating an appropriate vector system before genes can be transferred into a specific organism.

Animal Viruses

SV40 AND ADENOVIRUS

The first vectors developed for animal cells were derived from simple DNA viruses, which were relatively easy to manipulate by recombinant DNA techniques. Extra DNA, coding for foreign genes and for special markers ("tags") to track their progress, are inserted into the virus's chromosome. These passenger genes can be expressed via their own regulatory sequences or, sometimes more efficiently, via those of the virus.

The first animal virus used was SV40 (simian virus 40; Hamer et al., 1979; Mulligan et al., 1979). Fundamental studies on SV40 by Paul Berg and his coworkers laid the groundwork for their and other groups' subsequent development of it and other viruses as vectors for gene transfer, and earned Berg a Nobel Prize in 1980. SV40 can exist within the host cell both as an independent circular molecule or as a segment integrated in the host's DNA. This versatility, along with its well-characterized life cycle and gene regulation, have given researchers great flexibility in designing vector systems based on SV40. SV40's drawbacks are that it normally infects only cells of certain species (notably primates) and is severely limited in the amount of DNA it can carry. Only about 2,500 base pairs (the size of one small animal gene) can be added to this virus, and even this addition must be compensated for by deleting some of its own DNA.

Adenoviruses infect a wider variety of mammalian species than does SV40. Their DNA is a very long, linear molecule, which like SV40 can either replicate to give a high copy number of independent molecules or insert itself into the host's DNA in a low copy number. The molecular biology of adenoviruses has been well studied and like that of SV40, has provided fundamental insights into eucaryotic gene regulation.

Adenovirus vectors have several advantages over SV40 and retroviruses (which are discussed later). Adenovirus can accomodate large, complete passenger genes with their own control sequences. Furthermore, two different genes at widely separated locations can be accomodated on the same vector molecule, permitting separate and distinct control of the two passenger genes within one cell. In addition, hybrid viruses composed of both adenovirus and SV40 can give even greater flexibility in control of gene expression and extend the host range for gene transfer (van Doren and Gluzman, 1984).

Several developments with SV40 and adenoviruses are of particular interest. These viruses have been used to transfer genes into cells of diverse origin, notably mouse and human bone marrow cells (Karlsson et al., 1985). "Transient expression" with a recombinant SV40 vector was obtained at much higher frequency than with the calcium phosphate procedure. However, the recombinant SV40 vector did not integrate into the cells' chromosomes. With adenovirus-mediated transfer, one to three copies of foreign genes were transferred intact at very high frequency and maintained stably in the host cells' chromosomes. This low-copy number, stable integration is desirable for certain studies of gene regulation and for permanent genetic modification of animals.

Viral vectors can also be used for large-scale production of specific proteins in cultured animal cells. Although proteins can sometimes be efficiently manufactured in bacterial or yeast cells, many animal proteins are not correctly processed and assembled by cells of simpler organisms. In these cases it may be more efficient to manufacture proteins in cultured animal cells.

To be economically feasible, protein manufacture by recombinant DNA technology must yield large amounts of the desired product. Researchers have developed SV40 and adenovirus vectors that meet this requirement by expressing any inserted gene at a high level (Reddy et al., 1985; Yamada et al., 1985). The researchers made these "expression vectors" by connecting viral regulatory sequences that normally cause high-level production of proteins needed in huge quantities by the virus (e.g., coat proteins, which encase the thousands of viruses produced during infection of a cell) to genes for commercially desired proteins such as the hormone human choriogonadotropin, which is important in maintaining pregnancy. The expression vectors exploit the facts that many copies of viral DNA accumulate inside the cell and that each of these copies produces great quantities of the desired protein.

BOVINE PAPILLOMA VIRUS

Bovine papilloma virus (BPV) is another DNA virus under study and development as a vector for transferring mammalian genes (Sarver et al., 1981, 1982). This virus does not integrate its DNA into the host cell's chromosome. Instead, the vector with its passenger DNA is maintained as an extrachromosomal DNA molecule, which usually replicates to give about 100 copies of the transferred gene in every cell. The extrachromosomal maintenance and high copy number are advantageous for "transient expression" assays, detailed studies on gene expression, and production of proteins in quantity. An additional attribute is that BPV can carry large amounts of DNA—up to 20,000 base pairs.

The circular shape of BPV's DNA and its ability to maintain itself as an independent chromosome have enabled scientists to further engineer BPV (as well as SV40) vectors to replicate in both mammalian and bacterial cells. Researchers use these "shuttle vectors" to move cloned genes back and forth between mammalian and bacterial cells for ease of study and manipulation.

Drawbacks to BPV vectors are that the engineered DNA molecules are sometimes unstable, only a few types of cells (usually epithelial) can serve as hosts, and applications may be limited to cultured cells. Furthermore, in contrast to SV40, the basic biology of BPV is only now being characterized. Thus, researchers need to pursue fundamental studies on BPV's life cycle and regulatory mechanisms before optimal BPV vectors can be designed.

Vaccinia Virus

Vaccinia is a very large and complicated DNA virus. It is famous for its role as the vaccine used to eradicate the deadly human disease smallpox in this century. Although vaccinia is similar enough to the smallpox (variola) virus to immunize against it, vaccinia itself does not cause disease.

Poxviruses are unique in that they set up shop in the cell's cytoplasm, unlike other viruses, which head for the cell's nucleus. Vaccinia expresses its genes in the cytoplasm using its own enzymes, which respond to vaccinia's regulatory sequences but cannot recognize those of the host cell. Therefore, when vaccinia is used as a vector for foreign genes, these genes are expressed only if they are hooked up to vaccinia's own regulatory sequences.

Among its advantages is vaccinia's ability to grow easily in cell culture. By inoculation into the skin, it can also infect a wide range of animal hosts, making it a versatile vector. Moreover, similar poxviruses could be used as vectors for additional species. Because vaccinia is so large, it can accomodate more inserted DNA than any other virus—amounts greater than 25,000 base pairs are stable (Smith and Moss, 1983). This is more than 10 times the carrying capacity of SV40, and covers the size of several genes.

Vaccinia has two natural safety features: it does not integrate into its host's DNA, and it cannot become latent (i.e., persist in a dormant state for a long period). In addition, the virus can be attenuated further by genetic engineering. Scientists can insert passenger genes into the virus's gene for the enzyme thymidine kinase, thereby inactivating it. Because this enzyme is needed for optimal growth of the virus, vaccinia recombinants cannot spread as easily as the normal virus (Buller et al., 1985). In addition, viruses without thymidine kinase can survive treatment with a drug that kills the normal virus, enabling rapid laboratory detection of the desired recombinants.

Because the large vaccinia DNA molecule is too cumbersome to handle in vitro, foreign genes must be transferred onto the vaccinia vector by a two-step process. First a small circular "insertion vector" is built in vitro. This vector contains the foreign gene, surrounded by cloned DNA from vaccinia's thymidine kinase gene. Second, animal cells are infected with normal vaccinia virus, and then insertion vector DNA is added to the infected cells by direct DNA uptake. Inside the cells an exchange occurs between the thymidine kinase sequences on the insertion vector and the identical (homologous) sequences on the viral DNA, placing the foreign gene into the viral DNA. The foreign gene interrupts the thymidine kinase gene, inactivating it as described in the preceeding paragraph (Mackett et al., 1982).

The most important use of the vaccinia vector will be for the production of vaccines against viruses and parasites that have resisted conventional vaccines. Furthermore, a single recombinant vaccinia virus can carry antigenic genes from several disease agents or several strains of a virus like influenza. Thus vaccinia can immunize against several diseases in one shot (Perkus et al., 1985). Importantly, vaccinia vaccines not only stimulate antibody protection but also confer long-lasting cellular immunity (Bennink et al., 1984).

Recombinant vaccinia vaccines for major diseases of livestock (e.g., vesicular stomatitis virus, swine gastroenteritis) and for rabies, influenza, herpes simplex, hepatitis B, and some elements of malaria have already been successful in animal tests (Cremer et al., 1985; Mackett et al., 1985; Moss et al., 1984; Paoletti et al., 1984; Wiktor et al., 1984). Because of its wide host range, vaccinia can immunize a large variety of animal species. Like the original smallpox vaccine, the vaccines would be cheap, easy to manufacture, dispense, and administer, and stable without refrigeration as freeze-dried preparations—ideal for field use.

RETROVIRUSES

Retroviruses are a family of viruses that contain RNA as their primary genetic material. On infection of a host cell, the RNA is copied into DNA, which then inserts itself into the host cell's chromosome, becoming a stable part of the host's genetic information. Retroviruses have been found in association with many animals, including humans, and probably exist for all agriculturally important animal species.

There are several particular advantages to retroviral vectors (Anderson, 1984). They can infect a high percentage of the target cells, integrate in one copy at a single site in the cell's genome, and reliably express the foreign gene. Other methods often lead to the transfer of multiple copies of the gene, which may interfere with its correct expression.

Retroviruses are currently the focus of intense research on both their basic biology and their use as vectors. For example, engineered retroviruses can infect bone marrow cells in culture. These transformed cells can then be transplanted back into the animal. A gene introduced in this way may be able to correct a genetic defect in an animal or human, although it would not be inherited by the individual's progeny. However, infection of germ line cells of early embryos of animals should allow heritable traits to be transferred for breeding purposes in agriculture.

The first key experiments in the use of retroviral vectors concentrated on the transfer of genes for drug resistance into blood-producing cells of the mouse (Joyner et al., 1983; Williams et al., 1984) and of genes for the enzyme human hypoxanthine phosphoribosyltransferase (HPRT), whose absence causes Lesch-Nyhan syndrome, into mouse or human cells (Miller et al., 1983, 1984a; Willis et al., 1984). The HPRT gene functioned in both mouse and human cells in culture, as well as in live mice. Further experiments demonstrated efficient transfer of a rat growth hormone gene into mouse cells by retroviruses and correct expression of the gene by its own regulatory sequences (Miller et al., 1984b). More recently, β-globin genes were transferred into and correctly expressed in lines of transgenic mice (Soriano et al., 1986). This demonstrates that retroviruses can deliver genes into the germ cells of early embryos so that the genes are inherited normally and function in intact animals.

The engineering of safe retroviral vectors involves some genetic tricks to ensure that the virus will not be able to reinfect other cells or spread to other organisms after the desired transfer of genes. In constructing the vector some of the retrovirus's own genes are replaced with foreign passenger genes, depriving the virus of the ability to replicate itself. To overcome this handicap, a so-called "helper virus" is used, which provides gene products that the engineered retrovirus can no longer make. These essential products are the enzyme for replication and the proteins for the virus coat.

For the purpose of safety—and efficiency—the helper virus is debilitated by the removal of a small portion of the genetic material necessary to its reproduction. The helper is maintained only as an integrated "provirus" in a cell line; it is a permanent part of the cell's DNA and cannot become infectious. The handicapped vector retroviruses that carry foreign genes are propagated in this cell line, aided by the replication and coat proteins manufactured by the helper provirus. Vector viruses are then purified away from the cells containing the helper provirus.

These purified vectors now can enter other target cells and integrate the foreign gene into the target cells' genome, but that is all they can do—without the helper provirus they cannot replicate in the target cells to produce more infectious viruses. Thus the retroviral vector is a gene delivery system, not an infectious agent.

The vector can be further disabled by engineering a defective regulatory sequence at one end of its genome. Such vectors integrate into the host's chromosome, and then become stuck. Even in the presence of the helper virus, they cannot express their viral genes, replicate further, or move out of the cell's chromosome. Foreign genes transferred in by these vectors are expressed from their own regulatory sequences.

Retroviral gene transfer vectors applicable to agricultural animals have been developed. One system based on a turkey retrovirus efficiently delivers genes into avian and some mammalian cells (Watanabe and Temin, 1983). Another retrovirus system can introduce genes into a broad range of mammalian species, including farm animals (Cone and Mulligan, 1984). Thus, just a few retroviral vectors may serve for genetic engineering of many livestock species.

BACULOVIRUSES

Baculoviruses, which infect lepidopteran insects, should have uses in agriculture for manipulation of both beneficial and harmful species. They have already been used to express human β-interferon (Smith et al., 1983), c-*myc* protein (Miyamoto et al., 1985), interleukin 2 (Smith et al., 1985), and bacterial β-galactosidase (Pennock et al., 1984) in cultured insect cells, and human α-interferon in silkworm larvae (Maeda et al., 1985).

Baculoviruses have some similarities to vaccinia virus in the way they are engineered for gene transfer (Miller et al., 1986). Their large, double-stranded DNA genome may accomodate up to 100,000 extra base pairs of DNA, due to the virus's extendable rod-shaped structure. Insertion of genes into such a large DNA molecule is accomplished via small insertion vectors, as described previously for vaccinia. Viral and insertion vector DNA are simultaneously introduced into insect cells by direct uptake using calcium phosphate. Homologous recombination in vivo then places the foreign genes from the insertion vector into the baculovirus genome.

Foreign genes are most conveniently inserted into the virus's gene for polyhedrin. This strategy has several benefits. First, insertional inactivation of the polyhedrin gene gives an easily detected recombinant virus phenotype, because these viruses form areas of infected cells that look different from those made by the normal virus. Second, viruses with a defective polyhedrin gene cannot be transmitted between host insects; they can move only from cell to cell within a single insect or cell culture. Thus the recombinant baculoviruses have a built-in safety feature. Third, the regulatory sequence (promoter) of the polyhedrin gene can express foreign proteins at high levels, as over 20 percent of the infected cell's messenger RNA and protein are normally made from this gene. Foreign genes cloned in baculoviruses can also be expressed from their own promoters.

A baculovirus, high-level expression system could be used to manufacture commercially useful proteins, as baculoviruses can be mass-produced in insect cell cultures. Baculoviruses might be particularly advantageous for the manufacture of insect-derived substances such as pheromones, which can be used for biological control of insect pests.

Baculoviruses infect many lepidopteran insect species and can themselves be used as insecticides. Their effectiveness as biological insecticides may be augmented by genetic engineering, for example, by introduction of insect-specific toxin genes. Because baculoviruses infect only invertebrates, with different baculoviruses being relatively specific for certain lepidopteran insect hosts only, they should not spread indiscriminantly to other insects, animals, or plants.

Plant Viruses

CAULIFLOWER MOSAIC VIRUS

Only small steps have been taken with viral vectors for plants, in contrast to the great strides in virally mediated gene transfer into animals. There are no known plant retroviruses and only a few, small DNA viruses. The best-studied virus is cauliflower mosaic virus (CaMV), a small double-stranded DNA virus that infects cruciferous plants, such as cabbage and mustard. CaMV is transmitted in nature by aphids, but its DNA can infect plants if simply rubbed onto their leaves. CaMV causes systemic infection and replicates abundantly throughout the plant. It thus should transfer many copies of a gene per cell into all tissues of a mature plant. Furthermore, powerful CaMV gene regulation sequences can promote high-level expression of foreign genes. In fact, CaMV promoters are being used to augment the expression of plant genes transferred via other systems, as most plants recognize these promoters even when they are detached from the rest of CaMV.

The biggest obstacles to the development of a CaMV vector have been the severe limitation on the virus's size and thus on the quantity of DNA that can be inserted, and the instability of the genetically engineered virus. This instability may be caused both by the packaging limitation on extra DNA and by the way the virus replicates. Furthermore, CaMV does not integrate into plant genomes under normal conditions of infection. Some success in introducing foreign genes into plants using CaMV has been reported, however. Bacterial drug resistance genes were expressed and stably propagated in CaMV-infected turnip plants (Brisson et al., 1984).

GEMINIVIRUSES

Geminiviruses are single-stranded DNA viruses of plants that are transmitted by insects, such as leafhoppers. Viruses in this group infect many crops, including the monocots wheat and corn and the dicots beans, tobacco, and tomatoes. Work on developing a vector system based on these viruses is in progress (Kridl and Goodman, 1985; Lazarowitz, 1987).

Recently published experiments indicate that geminiviruses can be combined with the *Agrobacterium* Ti plasmid delivery system (described in a subsequent section) to obtain "agroinfection" of corn plants with the geminivirus maize streak virus (Grimsley et al., 1987). This dual system may prove useful in introducing engineered geminivirus vectors into plants, because often their DNA is not infectious unless transmitted as as intact virus by the natural insect mechanism. These experiments also demonstrated that *Agrobacterium* can transfer DNA to corn, a monocot, which was thought not to be amenable to the Ti plasmid gene transfer system.

RNA VIRUSES

Although there are many known plant RNA viruses, progress has been limited by the fact that manipulations developed to recombine DNA cannot be done on RNA directly. However, scientists can construct complementary DNA copies of RNA virus genomes. These copies can be used to construct a vector that will carry a foreign gene. The DNA can then be transcribed back into RNA, enabling the engineered virus to infect cells.

Brome mosaic virus (BMV), which infects monocots—including the important cereal crops—has been developed as a vector in this way by Ahlquist and coworkers (Ahlquist et al., 1984; Ahlquist and Janda, 1984; French et al., 1986). These researchers achieved transfer and expression of a bacterial drug resistance gene in barley protoplasts. The vector replicated rapidly within the cells, and the foreign gene, under the control of the powerful BMV promoter, was expressed at high levels within 20 hours of infection.

Although the plant remains infected with the virus, symptoms of infection vary greatly for different virus/host combinations. Sometimes symptoms are very mild: Wheat in some parts of the world is always infected with BMV to little effect, whereas infected barley suffers stunted growth. Desirable vectors would produce mild or no symptoms in the host plant and would not affect the plant's productivity in the field. These characteristics might be further improved by genetic engineering of the vector virus.

Control of viral disease, rather than introduction of a new trait, may be possible through exploitation of a natural phenomenon involving RNA viruses and their associated viral satellites. These are small nucleic acids that require the helper functions of a bona fide virus to replicate. They often attenuate the disease symptoms caused by that virus. Because satellites replicate rapidly at the expense of their helper, this molecular parasitism may provide a basis for viral disease control. Chinese scientists have placed RNA satellites in pepper plants in the field, but without gene transfer into the chromosome. The plants resisted viral infection (Tien and Chang, 1983).

Incorporation of satellite genes into the plant's chromosome could build in protection against disease symptoms caused by the helper virus (Kaper and Tousignant, 1984). To this end, British scientists have tranferred DNA copies of a viral satellite into the genome of tobacco plants, using the Ti plasmid system for gene transfer. The DNA copies functioned to produce satellite RNA (Baulcombe et al., 1986). Further testing is needed to determine whether the plants are resistant to viral disease. In a related development, American scientists engineered a single gene of the RNA virus tobacco mosaic virus (TMV) into tobacco and tomato plants via the Ti plasmid vector. Expression of this gene by the host plants made them resistant to infection by TMV (Abel et al., 1986). TMV causes large losses worldwide on cash crops such as tobacco, tomatoes, and bell peppers.

Viral satellites might also serve to transfer foreign genes into plants directly, in the manner described for BMV.

There are several potential advantages to RNA virus vector systems for plants. First, upon infection with cloned DNA or in vitro RNA transcripts, the plant should express the new trait immediately, in contrast to the Ti plasmid system (discussed in a subsequent section), in which a long regeneration process is usually necessary to obtain a transformed plant. Second, expression of the virus as an extrachromosomal, self-replicating RNA molecule means that gene expression will not be influenced by "position effects" due to insertion in undesirable places in the plant's chromosomes. Third, gene expression via the strong viral promoter, coupled with template replication to give many gene copies, would allow the production of large amounts of specific gene products within plant cells. Fourth, RNA viruses suitable to this strategy probably can be found for any host plant.

One possible problem with the use of RNA vectors, or any vector that replicates through an RNA intermediate (e.g., CaMV, retroviruses, and some transposons [see the next section]), is the high error rate associated with RNA replication. This might cause mutations detrimental to the foreign genes or to the vector itself during its replication cycle (van Vloten-Doting et al., 1985).

Transposable Elements

Transposable elements (also called "transposons") can move from place to place within an organism's genome and take extra pieces of DNA along for the ride. These elements have some physical and functional properties in common with retroviruses, but they do not spread from cell to cell by infection and therefore are not considered to be viruses. Barbara McClintock first recognized transposable elements in corn 40 years ago, for which she won a Nobel prize in 1983.

Transposable elements have since been found to be widespread in nature: examples have been described in bacteria, yeast, nematodes, fruit flies, mice, corn, soybeans, and snapdragons. It is likely that they will be found to exist for all species. Their apparent ubiquity in nature may make transposons especially useful for genetic modification of agronomically important insects and plants. Already, transposons have been used to modify *Pseudomonas fluorescens* bacteria that live on corn roots by insertion of an insecticidal gene from *Bacillus thuringiensis* (Obukowicz et al., 1986). In addition, studies of gene function aided by the use of transposable elements are very important for understanding basic aspects of gene expression in insects and plants. This knowledge is essential to the application of genetic engineering.

The jumping abilities of transposable elements have been used to isolate important genes from corn (Fedoroff et al., 1984). This is done by inducing the transposable element to jump into the corn gene of interest, thereby inactivating the gene and producing a mutant plant. When DNA from the mutant plant is compared to DNA from a normal plant, the characteristic sequence of the transposable element identifies its location, thus acting as a tag for the mutant gene. The DNA surrounding the transposable element is then cloned, yielding copies of the gene of interest. Although these copies are inactive because of the insertion of the transposable element, their sequences can be used as probes to find the active gene copy from a normal plant. This gene *isolation* strategy contrasts with gene *transfer* via transposable elements in that for isolation the transposable element is inserted into the gene, whereas for transfer the gene is inserted into the transposable element.

The major elements described in plants are Ac, Mu, and Spm in corn, Tgm1 in soybeans, and Tam1 in snapdragons. Researchers are trying to adapt these elements as vectors, particularly because they are so prevalent in the monocot crop corn, which has resisted most efforts to transfer genes via the most highly developed plant vector, the Ti plasmid. Engineered as vectors, transposable elements might be microinjected into corn embryos to transfer genes into the germ line, bypassing problems encountered with the introduction of DNA into cultured corn cells and the subsequent regeneration of plants.

The transposable P-element of the fruit fly *Drosophila melanogaster* has proved a very powerful tool for gene transfer in this organism (Rubin and Spradling, 1982; Spradling and Rubin, 1982). The principle of gene transfer is much the same as that described previously for retroviruses. The gene of interest is inserted into the P-element vector. This disrupts some functions of the P-element required for transposition, but these functions can be provided by a second, helper copy of the P-element. The helper has been engineered so that it cannot transpose itself but can still produce enzymes that cause transposition of the vector. Because transposable elements are not infectious in the way viruses are, both vector and helper P-elements must be microinjected into *Drosophila* embryos. The transposase enzyme of the helper acts upon DNA sequences located at the ends of the vector element, causing the vector to insert itself into the host's chromosomes. Large segments of DNA can be transferred in this manner. Transfer is efficient and stable and can be accomplished in the germ cells of the embryo, allowing the new trait to be inherited by future generations. Successful transfer, inheritance, and expression have been achieved with a wide variety of *Drosophila* genes. It may be possible to adapt the P-element or a similar system to other insects of agronomic importance.

The Ti Plasmid

The most successful gene transfer vector developed thus far for plant cells is the Ti plasmid found in the soil bacterium *Agrobacterium tumefaciens*. Plasmids are circular DNA molecules that exist independently of the cell's main chromosomes; the Ti plasmid is a naturally occurring variety that is quite large. *Agrobacterium* infects most species of dicots and causes a tumorous disease called crown gall. The disease is instigated by natural gene transfer of part of the bacterium's Ti plasmid, called T-DNA, into the plant's chromosomes. Plant cells acquire new properties as a consequence of the transferred genes. Besides metabolic changes that incite their uncontrolled growth into a tumor, the cells are programmed to manufacture certain chemical compounds called opines, which are used by the parasitic *Agrobacterium* as food. Thus *Agrobacterium tumefaciens* is a natural genetic engineer that forces a plant to do its bidding! It inserts its bacterial genes to create tumors composed of altered plant cells that provide it with specialized food.

Researchers have adapted the Ti plasmid to transfer foreign genes into plants and to obtain stable and heritable expression of the genes in normal, nontumorous plants. In order to be able to regenerate plants from cells transformed with T-DNA in culture, they modified the Ti plasmid to eliminate its tumor-promoting properties. Transferred genes can be expressed under the control of their own normal regulatory signals, or T-DNA signals can be used to turn on the foreign genes.

A strategy similar to that used for vaccinia is used to insert foreign genes within the T-DNA of the large Ti plasmid: transfer of engineered genes from a small plasmid insertion vector to the Ti plasmid by in vivo homologous DNA recombination within *Agrobacterium* cells. The T-DNA containing the foreign genes is then transferred from the Ti plasmid within *Agrobacterium* into the chromosomes of plant cells by its natural process (de Block et al., 1984; Fraley et al., 1983; Herrera-Estrella et al., 1983; Horsch et al., 1984, 1985).

Alternatively, a strategy like that of helper retroviruses or P-elements is used. In this case two separate plasmids are placed within *Agrobacterium*, one containing foreign genes cloned within the T-DNA's border sequences that enable the DNA segment to move, the other providing the helper functions that catalyze movement. Again, foreign genes contained between T-DNA border sequences are transferred into the plant cell (An et al., 1985; Bevan, 1984; Hoekema et al., 1983).

Plants currently amenable to Ti plasmid vectors include petunias, tobacco, soybeans, carrots, tomatoes, alfalfa, and oilseed rape. Genes transferred include the small subunit of the plant photosynthetic enzyme ribulose 1,5-bisphosphate carboxylase (Broglie et al., 1984; Herrera-Estrella et al., 1984), the bean storage protein phaseolin (Murai et al., 1983; Sengupta-Gopalan et al., 1985), the corn storage protein zein (Matzke et al., 1984), the wheat photosynthetic chlorophyll a/b binding protein (Lamppa et al., 1985), and a bacterial enzyme for resistance to the herbicide glyphosate (Comai et al., 1985).

Although many experiments focus on a basic understanding of plant gene regulatory mechanisms, experiments with herbicide and pest resistance genes are already introducing agronomic modifications into dicotyledonous crops such as soybeans, tomatoes, turnips, tobacco, and oilseed rape. Likewise, the nutritional improvement of seed crops is an important goal. One commercial firm is attempting to transfer the gene for a methionine- and cysteine-rich protein found in Brazil nuts to soybeans to improve their nutritional balance (Altenbach et al., in press).

Until recently it was thought that the Ti plasmid could not be used to transfer genes into monocots. This class of plants is not naturally infected by *Agrobacterium*. However, it now appears that at least some monocots can be transformed by DNA transferred from the Ti plasmid. T-DNA transfer and expression was demonstrated for asparagus and lilies (Hernalsteens et al., 1984; Hooykaas-van Slogteren et al., 1984), and more recently for corn (Graves and Goldman, 1986). In addition, the Ti vector can deliver DNA of a plant virus into corn plants (Grimsley et al., 1987). It will be an important breakthrough if the powerful Ti system can be usefully applied to the major monocot cereal crops corn, wheat, and rice.

An important factor in the use of the Ti plasmid system, as well as in direct DNA uptake and cell fusion methods, is the ability to regenerate whole plants from transformed cells. This still has not been accomplished for several major crops, but recent progress with rice is very encouraging (Abdullah et al., 1986; Fujimara et al., 1985; Yamada et al., 1986). Current efforts are also directed toward methods to introduce T-DNA into pollen grains, seeds, and seedlings, routes that bypass the steps of protoplast culture and regeneration.

Similar to the Ti plasmid, the Ri plasmid from *Agrobacterium rhizogenes* can be used to transfer genes into plants (David et al., 1984; Tepfer, 1984). This plasmid induces root proliferation in affected tissue. The roots are organized plant tissue, in contrast to Ti-induced tumors, which are masses of undifferentiated cells. The fast-growing root cultures are themselves useful for tests of new herbicides and pesticides developed to control pathogens that attack roots (Mugnier et al., 1986). Furthermore, the Ri plasmid vector can transfer new genes that confer resistance on the plant to herbicides, pesticides, or to the pathogens themselves.

Fungal and Bacterial Plasmids

Plasmids occur naturally in yeast, fungi, and bacteria. Scientists have used plasmid vectors extensively for basic research on the molecular biology of strains of these organisms commonly studied in the laboratory. With recombinant DNA techniques, researchers can cut and splice genes into small plasmids quite easily. Likewise, they can combine useful parts from different plasmids to create new plasmid vectors better suited to a particular gene transfer operation. Small plasmids can be introduced into cells by direct DNA uptake. Once inside the cell they replicate and stably maintain themselves and can express foreign genes that have been engineered into them. Furthermore, under certain conditions plasmids can transfer the foreign genes they carry into the host cells' chromosomes, where the genes can also be maintained and expressed. Thus plasmids are versatile vectors for gene transfer into procaryotes (bacteria) and simple eucaryotes (yeast and fungi).

Until recently, transformation systems were lacking for fungi of agricultural and industrial importance. For instance, the fungal corn pathogen *Cochliobolus heterostrophus* contains a toxin gene that might be manipulated to create a weed control agent or to develop resistant strains of corn. Progress was stymied, however, until the development of a plasmid-based gene transfer system for *C. heterostrophus* (Turgeon et al., 1985). Work on other pathogenic fungi is also progressing. The systems for pathogenic fungi rely on elements of a plasmid vector developed for the laboratory model fungus *Aspergillus nidulans* (Yelton et al., 1984).

Pathogenic and beneficial fungi and bacteria are important candidates for agronomically valuable gene transfer strategies. Pathogenic fungi and bacteria can be used as biological control agents for pest insects or weeds. Isolation and transfer of pathogenicity genes has a twofold purpose: construction of improved agents for biological control, and discovery of resistance genes in the plant or insect that counteract the pathogenicity. The use of transposons and plasmids to isolate and study pathogenic genes from fungi and bacteria that attack crop plants will lead to an understanding of the molecular bases of many agronomically critical diseases and suggest ways to combat them.

Beneficial fungi and bacteria may be improved and their host range extended to help other plants and animals. In addition, transformation of beneficial fungi and bacteria should prove advantageous for introducing improved traits for commercial production of special metabolites such as antibiotics and pigments and for food processing and waste disposal. For example, studies on bacteria with a natural capacity to degrade toxic herbicide and pesticide residues should yield improved strains that may prove useful in detoxifying the environment (Ghosal et al., 1985; Serdar and Gibson, 1985).

Another important aspect of bacterial gene transfer is basic and applied research on strains of *Rhizobium* that fix nitrogen for legumes. These studies have the following goals: improved strains of *Rhizobium*, engineered strains of other bacteria that can fix nitrogen for other crops such as cereals, and perhaps even crops that can fix nitrogen themselves. *Rhizobium* might also be used for the commercial production of ammonia.

Bacteria, as described at the outset, are generally easy targets for gene transfer. However, details must often be worked out for species that differ significantly from the laboratory model *Escherichia coli.*

An example of bacterial gene transfer for agricultural purposes is the transfer of an insecticidal toxin gene from *Bacillus thuringiensis* to a *Pseudomonas fluorescens* strain that colonizes corn roots, to extend the number of plant hosts that can be protected against pest insects by the bacterial toxin. *B. thuringiensis* itself has been marketed as an insecticide for many years. After ingestion, its toxin is activated in the insect's gut. There are different strains of *B. thuringiensis* that make toxins capable of killing over 100 different lepidopteran and dipteran pests. These toxins are harmless except to targeted insects, and delivery via bacteria with a specific range of plant hosts ensures a high level of specificity for the pesticide.

Scientists at Monsanto Company have transferred the *B. thuringiensis* toxin gene into *P. fluorescens* via a plasmid and also into the *P. fluorescens* chromosome via a transposon (Obukowicz et al., 1986; Watrud et al., 1985). The new biological insecticide is intended to protect corn against the black cutworm. *P. fluorescens* does not persist in the field, so the genetically engineered bacteria should kill off insects after application early in the growing season and then die.

A second strategy is to transfer the toxin gene into crops, to make them self-protecting. Scientists at the Belgian company Plant Genetic Systems engineered the *B. thuringiensis* toxin gene into plants, which then expressed the toxin and resisted insect predators (Vaeck et al., 1987).

A novel vector for introduction of genes into plants may result from studies on corynebacteria. These microbes colonize grasses, including wheat, corn, and sorghum. Some species are pathogenic to plants, others are harmless. Corynebacteria have their own plasmids, into which foreign genes could be inserted, and their proteins produced by the corynebacteria could easily be secreted through bacterial cell walls. Thus, foreign proteins expressed within corynebacteria might be made readily available to the plant host. Candidate products include insecticides, herbicides, antibiotics, and growth regulators. Japanese researchers have transformed certain food strains of corynebacteria and intend to use genetic engineering to improve their commercial production of amino acids. Transformation of field strains has proved more difficult but should soon be feasible (A. Vidaver, personal communication).

PROSPECTS

Molecular biologists have made tremendous strides since the early 1970s in experimental gene transfer and expression. Technology and the knowledge on which it is based continue to advance rapidly. It is truly remarkable that within 15 years, gene transfer has evolved from an esoteric technique practiced by a few bacterial and viral geneticists to a popular procedure that researchers in disparate biological fields use for wide-ranging studies.

This review has described major gene transfer methods with immediate potential for agricultural research. Diverse techniques are available: direct uptake of DNA, microinjection of DNA, cell fusion, and gene delivery by an array of vectors. Although details differ among animals, plants, and bacteria, underlying principles do not. Thus progress with one organism may have application to other systems by analogy. A recent example is electroporation, direct DNA transfer in a highly charged electric field. First achieved in 1982 with cultured mammalian cells, it has been widely adopted and further adapted for plant cells. Likewise, general strategies for structuring and manipulating vectors may be applied to many organisms. For instance, helper viruses that contribute essential life-support functions for defective viruses are a paradigm now adapted for gene transfer vectors derived from viruses, transposons, and plasmids.

There are still important problems that must be solved, however, in order to design optimal gene transfer systems. Molecular biologists still lack knowledge about many detailed mechanisms governing DNA uptake, integration into chromosomes, and gene regulation. Current approaches for DNA transfer therefore rely largely on experience and observation. They might be vastly improved by a more thorough understanding of the underlying molecular mechanisms.

The most serious problem for modification of animals and plants is that of obtaining correctly regulated gene expression in the appropriate tissues of the target organism. Regulation and stability of introduced genes is unfortunately still variable, although much progress is being made. For example, papers have reported correct tissue-specific and developmental regulation of a globin gene (for the blood protein hemoglobin) in transgenic mice (Chada et al., 1985; Magram et al., 1985). The transferred globin gene was turned on for the first time in fetal blood-forming cells, and was expressed in adult mice only in the blood-forming tissues bone marrow and spleen. This pattern mimics normal regulation of the mouse globin gene.

However, scientists have not yet devised a way to get genes to integrate at a specific site in animal or higher plant cell chromosomes, which would help in obtaining correct regulation and stability of introduced genes. Furthermore, with some current techniques genes are frequently inserted into chromosomes as multiple copies, confounding these problems. DNA rearrangements, lethal insertions, male-sterile mutations, and mosaic organisms in which not all cells contain the new gene sometimes result. Moreover, changes in a gene's expression sometimes occur after transmission of the new gene to progeny.

Researchers are working toward the ideal of targeted insertion of one stable gene copy that will be sexually transmitted and correctly expressed in all progeny. Fundamental studies of DNA recombination in mammalian cells that may lead to targeted integration are being carried out (Shaul et al., 1985; Thomas and Capecchi, 1986). Another strategy to obtain correct gene regulation, for instance of globin, is to insert a very large chromosomal segment that contains the gene surrounded by its usual neighboring genes. Genes within such a cluster may be correctly regulated by complex sequences in the surrounding DNA. Currently, very large segments of DNA are difficult to handle and require special vectors to accomodate them.

In most cases, researchers wish to keep inserted genes silent during early embryogenesis, and then activate them at the appropriate time in the organism's development. However, inserted genes are not always controlled correctly, even when their own regulatory sequences are still attached to them. For example, an inserted growth hormone gene controlled by its own promoter was not regulated correctly in transgenic mice, causing the female mice to be sterile. In contrast, when the same growth hormone gene was attached to the promoter from another gene, metallothionein, it could be turned on or off by raising or lowering the amount of trace metals in the transgenic animals' diets. Researchers could thus control both the amount of growth hormone made by transgenic animals and the time at which it was made.

Transfer of genes into cultured plant cells, from which transgenic plants will be regenerated, presents an additional problem—the traits targeted for the mature plants may not be expressed in cultured cells and conversely, successful expression in cultured cells may not carry over to plants regenerated from these cells. For example, a salt-tolerant cell in culture may not yield a salt-tolerant plant when regenerated. The problem of selecting traits in cell culture extends to all plant gene transfer techniques performed on cultured cells, including direct DNA uptake, DNA microinjection, cell fusion, and vector-mediated methods. Scientists are therefore working on adapting existing systems to deliver genes into pollen grains, seeds, and seedlings, which can develop normally into mature plants. This strategy has the additional advantage of obviating the need for plant cell culture and regeneration techniques for each individual species, which have been stumbling blocks for gene transfer into some agronomically important plants, notably the monocots corn and wheat, with recent progress being made for rice. Development of vectors for these species has also lagged behind, although adaptation of the Ti plasmid used for dicots, or transposons from monocots, may prove feasible. Alternatively, direct DNA uptake or microinjection of pollen or embryos might be used.

The problems of uptake and subsequent localization of DNA still impede research with some organisms, although these problems are being overcome, for example, in the pathogenic fungi. These problems extend also to compartments of eucaryotic cells other than the nucleus that contain their own DNA—mitochondria (the cell's energy powerhouses) and chloroplasts of green plants (which harness the energy of sunlight through the process of photosynthesis). For instance, the chloroplast's DNA encodes proteins essential to photosynthesis, and often related to these, proteins involved in herbicide resistance. An important goal yet to be achieved is the directed transport of new DNA into the plant chloroplast, although there is some experimental evidence to suggest that the Ti plasmid might be used (de Block et al., 1985). Other possible methods include microinjection of DNA directly into chloroplasts and introduction of new genes on plasmids that would be stably maintained in chloroplasts.

Additional basic and applied research is needed to extend existing gene transfer systems to agriculturally important organisms. Important practical details cannot always be extrapolated from well-studied laboratory models. Furthermore, scientists still lack basic biochemical and genetic knowledge about many agriculturally important species. This knowledge base is necessary to support more applied goals.

Gene transfer systems require a supply of agriculturally useful genes, if such systems are to benefit the farming community and other segments of society. Scientists must devise ways to find and isolate genes of agricultural interest. This can often be facilitated by the very gene transfer methods that will later be used to move the genes so identified into new hosts. Scientists must also devise methods to measure the presence of genes that are not easily detected immediately after transfer. It should also be noted that current methods are applicable only to dominant or co-dominant genes, since transfer of a recessive gene cannot change a trait within an organism unless the normal, dominant gene can be inactivated.

In summary, a variety of gene transfer methods is needed to accomplish diverse goals, which include fundamental studies of gene regulation, isolation of genes whose function and location are unknown, production of proteins in large quanities, and introduction of new traits.

REFERENCES

Abdullah, R., E. C. Cocking, and J. A. Thompson. 1986. Efficient plant regeneration from rice protoplasts through somatic embryogenesis. Bio/Technology 4:1087–1090.

Abel, P. P., R. S. Nelson, B. De, N. Hoffmann, S. G. Rogers, R. T. Fraley, and R. N. Beachy. 1986. Delay of disease development in transgenic plants that express the tobacco mosaic virus coat protein gene. Science 232:738–743.

Ahlquist, P., R. French, M. Janda, and L. S. Loesch-Fries. 1984. Multicomponent RNA plant virus infection derived from cloned viral cDNA. Proc. Natl. Acad. Sci. USA 81:7066–7070.

Ahlquist, P., and M. Janda. 1984. cDNA cloning and in vitro transcription of the complete brome mosaic virus genome. Mol. Cell. Biol. 4:2876–2882.

An, G., B. D. Watson, S. Stachel, M. P. Gordon, and E. W. Nester. 1985. New cloning vehicles for transformation of higher plants. EMBO J. 4:277–284.

Anderson, W. F. 1984. Prospects for human gene therapy. Science 226:401–409.

Altenbach, S. B., K. W. Pearson, F. W. Leung, and S. M. Sun. In press. Cloning and sequence analysis of a cDNA encoding a Brazil nut protein exceptionally rich in methionine. Plant Mol. Biol. Rep.

Austin, S., M. A. Baer, and J. P. Helgeson. 1985. Transfer of resistance to potato leaf roll virus from *Solanum brevidens* into *Solanum tuberosum* by somatic fusion. Plant Sci. 39:75–82.

Baulcombe, D. C., G. R. Saunders, M. W. Bevan, M. A. Mayo, and B. D. Harrison. 1986. Expression of biologically active viral satellite RNA from the nuclear genome of transformed plants. Nature 321:446–449.

Bennink, J. R., J. W. Yewdell, G. L. Smith, C. Moller, and B. Moss. 1984. Recombinant vaccinia virus primes and stimulates influenza haemagglutinin-specific cytotoxic T cells. Nature 311:578–579.

Benvenisty, N., and L. Reshef. 1986. Direct introduction of genes into rats and expression of the genes. Proc. Natl. Acad. Sci. USA 83:9551–9555.

Bevan, M. 1984. Binary *Agrobacterium* vectors for plant transformation. Nucleic Acids Res. 12:8711–8721.

Bravo, J. E., and D. A. Evans. 1985. Protoplast fusion for crop improvement. Plant Breeding Rev. 3:193–218.

Brinster, R. L., K. A. Ritchie, R. E. Hammer, R. L. O'Brien, B. Arp, and U. Storb. 1983. Expression of a microinjected immunoglobulin gene in the spleen of transgenic mice. Nature 306:332–336.

Brisson, N., J. Paszkowski, J. R. Penswick, B. Gronenborn, I. Potrykus, and T. Hohn. 1984. Expression of a bacterial gene in plants by using a viral vector. Nature 310:511–514.

Broglie, R., G. Coruzzi, R. T. Fraley, S. G. Rogers, R. B. Horsch, J. G. Niedermeyer, C. L. Fink, J. S. Flick, and N.-H. Chua. 1984. Light-regulated expression of a pea ribulose-1,5-bisphosphate carboxylase small subunit gene in transformed plant cells. Science 224:838–843.

Buller, R. M. L., G. L. Smith, K. Cremer, A. L. Notkins, and B. Moss. 1985. Decreased virulence of recombinant vaccinia virus expression vectors is associated with a thymidine kinase-negative phenotype. Nature 317:813–815.

Chada, K., J. Magram, K. Raphael, G. Radice, E. Lacy, and F. Costantini. 1985. Specific expression of a foreign β-globin gene in erythroid cells of transgenic mice. Nature 314:377–380.

Chourrout, D., R. Guyomard, and L. M. Houdebine. 1986. High efficiency gene transfer in rainbow trout (*Salmo gairdneri* Rich.) by microinjection into egg cytoplasm. Aquaculture 51:143–150.

Comai, L., D. Facciotti, W. R. Hiatt, G. Thompson, R. E. Rose, and D. M. Stalker. 1985. Expression in plants of a mutant *aro*A gene from *Salmonella typhimurium* confers tolerance to glyphosate. Nature 317:741–744.

Cone, R. D., and R. C. Mulligan. 1984. High-efficiency gene transfer into mammalian cells: generation of helper-free recombinant retrovirus with broad mammalian host range. Proc. Natl. Acad. Sci. USA 81:6349–6353.

Cremer, K. J., M. Mackett, C. Wohlenberg, A. L. Notkins, and B. Moss. 1985. Vaccinia virus recombinant expressing herpes simplex virus type 1 glycoprotein D prevents latent herpes in mice. Science 228:737–740.

Crossway, A., H. Hauptli, C. M. Houck, J. M. Irvine, J. V. Oakes, and L. A. Perani. 1986. Micromanipulation techniques in plant biotechnology. BioTechniques 4:320–334.

David, C., M.-D. Chilton, and J. Tempé. 1984. Conservation of T-DNA in plants regenerated from hairy root cultures. Bio/Technology 2:73–76.

de Block, M., L. Herrera-Estrella, M. van Montagu, J. Schell, and P. Zambryski. 1984. Expression of foreign genes in regenerated plants and in their progeny. EMBO J. 3:1681–1689.

de Block, M., J. Schell, and M. van Montagu. 1985. Chloroplast transformation by *Agrobacterium tumefaciens*. EMBO J. 4:1367–1372.

de la Peña, A., H. Lörz, and J. Schell. 1987. Transgenic rye plants obtained by injecting DNA into young floral tillers. Nature 325:274–276.

Dubensky, T. W., B. A. Campbell, and L. P. Villarreal. 1984. Direct transfection of viral and plasmid DNA into the liver or spleen of mice. Proc. Natl. Acad. Sci. USA 81:7529–7533.

Evans, D. A., C. E. Flick, and R. A. Jensen. 1981. Somatic hybrid plants between sexually incompatible species of the genus *Nicotiana*. Science 213:907–909.

Fedoroff, N. D., D. Furtek, and O. Nelson. 1984. Cloning of the *Bronze* locus in maize by a simple and generalizable procedure using the transposable controlling element *Ac*. Proc. Natl. Acad. Sci. USA 81:3825–3829.

Fraley, R. T., S. G. Rogers, R. B. Horsch, P. R. Sanders, J. S. Flick, S. P. Adams, M. L. Bittner, L. A. Brand, C. L. Fink, J. S. Fry, G. R. Galluppi, S. B. Goldberg, N. L. Hoffmann, and S. C. Woo. 1983. Expression of bacterial genes in plant cells. Proc. Natl. Acad. Sci. USA 80:4803–4807.

Frels, W. I., J. A. Bluestone, R. J. Hodes, M. R. Capecchi, and D. S. Singer. 1985. Expression of a microinjected porcine class I major histocompatibility complex gene in transgenic mice. Science 228:577–580.

French, R., M. Janda, and P. Ahlquist. 1986. Bacterial gene inserted in an engineered RNA virus: efficient expression in monocotyledonous plant cells. Science 231:1294–1297.

Fries, R., and F. H. Ruddle. 1986. Gene mapping in domestic animals. In Biotechnology for Solving Agricultural Problems (pp. 19–37), P. C. Augustine, H. D. Danforth, and M. R. Bakst, eds. Dordrecht, the Netherlands: Martinus Nijhoff.

Fromm, M., L. P. Taylor, and V. Walbot. 1985. Expression of genes transferred into monocot and dicot plant cells by electroporation. Proc. Natl. Acad. Sci. USA 82:5824–5828.

Fromm, M. E., L. P. Taylor, and V. Walbot. 1986. Stable transformation of maize after gene transfer by electroporation. Nature 319:791–793.

Fujimura, T., M. Sakurai, H. Akagi, T. Negishi, and A. Hirose. 1985. Regeneration of rice plants from protoplasts. Plant Tissue Culture Letters 2:74–75.

Gamble, H. R. 1986. Applications of hybridoma technology to problems in the agricultural sciences. In Biotechnology for Solving Agricultural Problems (pp. 39–52), P. C. Augustine, H. D. Danforth, and M. R. Bakst, eds. Dordrecht, the Netherlands: Martinus Nijhoff.

Gill, J. A., J. P. Sumpter, E. M. Donaldson, H. M. Dye, L. Souza, T. Berg, J. Wypych, and K. Langley. 1985. Recombinant chicken and bovine growth hormones accelerate growth in aquacultured juvenile pacific salmon *Oncorhynchus kisutch.* Bio/Technology 3:643–646.

Ghosal, D., I.-S. You, D. K. Chatterjee, and A. M. Chakrabarty. 1985. Microbial degradation of halogenated compounds. Science 228:135–142.

Graham, F. L., and A. J. van der Eb. 1973. A new technique for the assay of infectivity of human adenovirus 5 DNA. Virology 52:456–467.

Graves, A. C. F., and S. L. Goldman. 1986. The transformation of *Zea mays* seedlings with *Agrobacterium tumefaciens*—Detection of T-DNA specific enzyme activities. Plant Mol. Biol. Rep. 7:43–50.

Greisbach, R. J. 1983. Protoplast microinjection. Plant Mol. Biol. Rep. 1:32–37.

Greisbach, R. J. 1987. Chromosome-mediated transformation via microinjection. Plant Sci. In press.

Grimsley, N., T. Hohn, J. W. Davies, and B. Hohn. 1987. *Agrobacterium*-mediated delivery of infectious maize streak virus into maize plants. Nature 325:177–179.

Hamer, D. H., K. D. Smith, S. H. Boyer, and P. Leder. 1979. SV40 recombinants carrying rabbit β-globin gene coding sequences. Cell 17:725–735.

Hammer, R. E., R. D. Palmiter, and R. L. Brinster. 1984. Partial correction of murine hereditary growth disorder by germ-line incorporation of a new gene. Nature 311:65–69.

Hammer, R. E., V. G. Pursel, C. E. Rexroad, Jr., R. J. Wall, D. J. Bolt, K. M. Ebert, R. D. Palmiter, and R. L. Brinster. 1985. Production of transgenic rabbits, sheep and pigs by microinjection. Nature 315:680–683.

Hernalsteens, J.-P., L. Thia-Toong, J. Schell, and M. van Montagu. 1984. An *Agrobacterium*-transformed cell culture from the monocot *Asparagus officinalis.* EMBO J. 3:3039–3041.

Herrera-Estrella, L., M. de Block, E. Messens, J.-P. Hernalsteens, M. van Montagu, and J. Schell. 1983. Chimeric genes as dominant selectable markers in plant cells. EMBO J. 2:987–995.

Herrera-Estrella, L., G. van den Broeck, R. Maenhaut, M. van Montagu, J. Schell, M. Timko, and A. Cashmore. 1984. Light inducible and chloroplast-associated expression of a chimeric gene introduced into *Nicotiana tabacum* using a Ti plasmid vector. Nature 310:115–120.

Hoekema, A., P. R. Hirsch, P. J. J. Hooykaas, and R. A. Schilperoort. 1983. A binary plant vector strategy based on separation of *vir*- and T-region of the *Agrobacterium tumefaciens* Ti-plasmid. Nature 303:179–181.

Hooykaas-van Slogteren, G. M. S., P. J. J. Hooykaas, and R. A. Schilperoort. 1984. Expression of Ti plasmid genes in monocotyledonous plants infected with *Agrobacterium tumefaciens.* Nature 311:763–764.

Horsch, R. B., R. T. Fraley. S. G. Rogers, P. R. Sanders, A. Lloyd, and N. Hoffmann. 1984. Inheritance of functional foreign genes in plants. Science 223:496–498.

Horsch, R. B., J. E. Fry, N. L. Hoffmann, D. Eichholtz, S. G. Rogers, and R. T. Fraley. 1985. A simple and general method for transferring genes into plants. Science 227:1229–1231.

Joyner, A., G. Keller, R. A. Phillips, and A. Bernstein. 1983. Retrovirus transfer of a bacterial gene into mouse haematopoietic progenitor cells. Nature 305:556–558.

Kaper, J. M., and M. E. Tousignant. 1984. Viral satellites: parasitic nucleic acids capable of modulating disease expression. Endeavour, New Series 8:194–200.

Karlsson, S., R. K. Humphries, Y. Gluzman, and A. W. Nienhuis. 1985. Transfer of genes into hematopoietic cells using recombinant DNA viruses. Proc. Natl. Acad. Sci. USA 82:158–162.

Kridl, J. C., and R. M. Goodman. 1986. Transcriptional regulatory sequences from plant viruses. BioEssays 4:4–8.

Lamppa, G., F. Nagy, and N.-H. Chua. 1985. Light-regulated and organ-specific expression of a wheat Cab gene in transgenic tobacco. Nature 316:750–752.

Lazarowitz, S. G. 1987. The molecular characterization of geminiviruses. Plant Mol. Biol. Rep. In press.

Lörz, H., B. Baker, and J. Schell. 1985. Gene transfer to cereal cells mediated by protoplast transformation. Mol. Gen. Genet. 199:178–182.

Mackett, M., G. L. Smith, and B. Moss. 1982. Vaccinia virus: a selectable eukaryotic cloning and expression vector. Proc. Natl. Acad. Sci. USA 79:7415–7419.

Mackett, M., T. Yilma, J. K. Rose, and B. Moss. 1985. Vaccinia virus recombinants: expression of VSV genes and protective immunization of mice and cattle. Science 227:433–435.

Maeda, S., T. Kawai, M. Obinata, H. Fujiwara, T. Horiuchi, Y. Saeki, Y. Sato, and M. Furusawa. 1985. Production of human α-interferon in silkworm using a baculovirus vector. Nature 315:592–594.

Magram, J., K. Chada, and F. Costantini. 1985. Developmental regulation of a cloned adult β-globin gene in transgenic mice. Nature 315:338–340.

Matzke, M. A., M. Susani, A. N. Binns, E. D. Lewis, I. Rubenstein, and A. J. M. Matzke. 1984. Transcription of a zein gene introduced into sunflower using a Ti plasmid vector. EMBO J. 3:1525–1531.

McCutchan, J. H., and J. S. Pagano. 1968. Enhancement of the infectivity of simian virus 40 deoxyribonucleic acid with diethylaminoethyl-dextran. J. Natl. Cancer Inst. 41:351–357.

McKnight, G. S., R. E. Hammer, E. A. Kuenzel, and R. L. Brinster. 1983. Expression of the chicken transferrin gene in transgenic mice. Cell 34:335–341.

Miller, A. D., R. J. Eckner, D. J. Jolly, T. Friedmann, and I. M. Verma. 1984a. Expression of a retrovirus encoding human HPRT in mice. Science 225:630–632.

Miller, A. D., D. J. Jolly, T. Friedmann, and I. M. Verma. 1983. A transmissible retrovirus expressing human hypoxanthine phosphoribosyltransferase (HPRT): gene transfer into cells obtained from humans deficient in HPRT. Proc. Natl. Acad. Sci. USA 80:4709–4713.

Miller, A. D., E. S. Ong, M. G. Rosenfeld, I. M. Verma, and R. M. Evans. 1984b. Infectious and selectable retrovirus containing an inducible rat growth hormone minigene. Science 225:993–998.

Miller, D. W., P. Safer, and L. K. Miller. 1986. An insect baculovirus host-vector system for high-level expression of foreign genes. In Genetic Engineering, Vol. 8 (pp. 277–298), J. K. Setlow and A. Hollaender, eds. New York: Plenum Press.

Miyamoto, C., G. E. Smith, J. Farrell-Towt, R. Chizzonite, M. D. Summers, and G. Ju. 1985. Production of human c-*myc* protein in insect cells infected with a baculovirus expression vector. Mol. Cell. Biol. 5:2860–2865.

Moss, B., G. L. Smith, J. L. Gerin, and R. H. Purcell. 1984. Live recombinant vaccinia virus protects chimpanzees against hepatitis B. Nature 311:67–69.

Mugnier, J., P. W. Ready, and G. E. Riedel. 1986. Root culture system useful in the study of biotrophic root pathogens in vitro. In Biotechnology for Solving Agricultural Problems (pp. 147–153), P. C. Augustine, H. D. Danforth, and M. R. Bakst, eds. Dordrecht, the Netherlands: Martinus Nijhoff.

Mulligan, R. C., B. H. Howard, and P. Berg. 1979. Synthesis of rabbit β-globin in cultured monkey kidney cells following infection with a SV40 β-globin recombinant genome. Nature 277:108–114.

Murai, N., D. W. Sutton, M. G. Murray, J. L. Slighton, D. J. Merlo, N. A. Reichert, C. Sengupta-Gopalan, C. A. Stock, R. F. Barker, J. D. Kemp, and T. C. Hall. 1983. Phaseolin gene from bean is expressed after transfer to sunflower via tumor-inducing plasmid vectors. Science 222:476–482.

Neumann, E., M. Schaefer-Ridder, Y. Wang, and P. H. Hofschneider. 1982. Gene transfer into mouse lyoma cells by electroporation in high electric fields. EMBO J. 1:841–845.

Obukowicz, M. G., F. J. Perlak, K. Kusano-Kretzmer, E. J. Mayer, S. L. Bolten, and L. S. Watrud. 1986. Tn5-mediated integration of the Delta-endotoxin gene from *Bacillus thuringiensis* into the chromosome of root-colonizing pseudomonads. J. Bacteriol. 168:982–989.

Palmiter, R. D., G. Norstedt, R. E. Gelinas, R. E. Hammer, and R. L. Brinster. 1983. Metallothionein-human GH fusion genes stimulate growth of mice. Science 222:809–814.

Paoletti, E., B. R. Lipinskas, C. Samsonoff, S. Mercer, and D. Panicali. 1984. Construction of live vaccines using genetically engineered poxviruses: biological activity of vaccinia virus recombinants expressing the hepatitis B virus surface antigen and the herpes simplex virus glycoprotein D. Proc. Natl. Acad. Sci. USA 81:193–197.

Pelletier, G., C. Primard, F. Vedel, and P. Chetrit. 1983. Intergeneric cytoplasmic hybridization in Cruciferae by protoplast fusion. Mol. Gen. Genet. 191:244–250.

Pennock, G. D., C. Shoemaker, and L. K. Miller. 1984. Strong and regulated expression of *Escherichia coli* β-galactosidase in insect cells with a baculovirus vector. Mol. Cell. Biol. 4:399–406.

Perkus, M. E., A. Piccini, B. R. Lipinskas, and E. Paoletti. 1985. Recombinant vaccinia virus: immunization against multiple pathogens. Science 229:981–984.

Potrykus, I., M. Saul, J. Petruska, J. Paszkowski, and R. D. Shillito. 1985a. Direct gene transfer to cells of a graminaceous monocot. Mol. Gen. Genet. 199:183–188.

Potrykus, I., R. D. Shillito, M. W. Saul, and J. Paszkowski. 1985b. Direct gene transfer—state of the art and future potential. Plant Mol. Biol. Rep. 3:117–128.

Potter, H., L. Weir, and P. Leder. 1984. Enhancer-dependent expression of human κ immunoglobulin genes introduced into mouse pre-B lymphocytes by electroporation. Proc. Natl. Acad. Sci. USA 81:7161–7165.

Reddy, V. B., A. K. Beck, A. J. Garramone, V. Vellucci, J. Lustbader, and E. G. Bernstine. 1985. Expression of human choriogonadotropin in monkey cells using a single simian virus 40 vector. Proc. Natl. Acad. Sci. USA 82:3644–3648.

Rubin, G. M., and A. C. Spradling. 1982. Genetic transformation of *Drosophila* with transposable element vectors. Science 218:348–353.

Sarver, N., J. C. Byrne, and P. M. Howley. 1982. Transformation and replication in mouse cells of a bovine papillomavirus-pML2 plasmid vector that can be rescued in bacteria. Proc. Natl. Acad. Sci. USA 79:7147–7151.

Sarver, N., P. Gruss, M.-F. Law, G. Khoury, and P. M. Howley. 1981. Bovine papilloma virus deoxyribonucleic acid: a novel eucaryotic cloning vector. Mol. Cell. Biol. 1:486–496.

Schocher, R. J., R. D. Shillito, M. W. Saul, J. Paszkowski, and I. Potrykus. 1986. Co-transformation of unlinked foreign genes into plants by direct gene transfer. Bio/Technology 4:1093–1096.

Sekine, S., T. Mizukami, T. Nishi, Y. Kuwana, A. Saito, M. Sato, S. Itoh, and H. Kawauchi. 1985. Cloning and expression of cDNA for salmon growth hormone in *Escherichia coli*. Proc. Natl. Acad. Sci. USA 82:4306–4310.

Sengupta-Gopalan, C., N. A. Reichert, R. F. Barker, T. C. Hall, and J. D. Kemp. 1985. Developmentally regulated expression of the bean β-phaseolin gene in tobacco seed. Proc. Natl. Acad. Sci. USA 82:3320–3324.

Serdar, C. M., and D. T. Gibson. 1985. Enzymatic hydrolysis of organophosphates: cloning and expression of a parathion hydrolase gene from *Pseudomonas diminuta*. Bio/Technology 3:567–571.

Shani, M. 1985. Tissue-specific expression of rat myosin light-chain 2 gene in transgenic mice. Nature 314:283–286.

Shaul, Y., O. Laub, M. D. Walker, and W. J. Rutter. 1985. Homologous recombination between a defective virus and a chromosomal sequence in mammalian cells. Proc. Natl. Acad. Sci. USA 82:3781–3784.

Shillito, R. D., M. W. Saul, J. Paszkowski, M. Müller, and I. Potrykus. 1985. High efficiency direct gene transfer to plants. Bio/Technology 3:1099–1103.

Smith, G. E., G. Ju, B. L. Ericson, J. Moschera, H.-W. Lahm, R. Chizzonite, and M. D. Summers. 1985. Modification and secretion of human interleukin 2 produced in insect cells by a baculovirus expression vector. Proc. Natl. Acad. Sci. USA 82:8404–8408.

Smith, G. E., M. D. Summers, and M. J. Fraser. 1983. Production of human beta interferon in insect cells infected with a baculovirus expression vector. Mol. Cell. Biol. 3:2156–2165.

Smith, G. L., and B. Moss. 1983. Infectious poxvirus vectors have capacity for at least 25,000 base pairs of foreign DNA. Gene 25:21–28.

Soriano, P., R. D. Cone, R. C. Mulligan, and R. Jaenisch. 1986. Tissue-specific and ectopic expression of genes introduced into transgenic mice by retroviruses. Science 234:1409–1413.

Spradling, A. C., and G. M. Rubin. 1982. Transposition of cloned P elements into *Drosophila* germ line chromosomes. Science 218:341–347.

Swift, G. H., R. E. Hammer, R. J. MacDonald, and R. L. Brinster. 1984. Tissue-specific expression of the rat pancreatic elastase I gene in transgenic mice. Cell 38:639–646.

Tepfer, D. 1984. Transformation of several species of higher plants by *Agrobacterium rhizogenes*: sexual transmission of the transformed genotype and phenotype. Cell 37:959–967.

Thomas, K. R., and M. R. Capecchi. 1986. Introduction of homologous DNA sequences into mammalian cells induces mutations in the cognate gene. Nature 324:34–38.

Tien, P., and X. H. Chang. 1983. Control of two seed-borne virus diseases in China by the use of protective inoculation. Seed Sci. Technol. 11:969–972.

Turgeon, B. G., R. C. Garber, and O. C. Yoder. 1985. Transformation of the fungal maize pathogen *Cochliobolus heterostrophus* using the *Aspergillus nidulans amdS* gene. Mol. Gen. Genet. 201:450–453.

Vaeck, M., A. Reynaerts, H. Hofte, M. van Montagu, and J. Leemans. 1987. New developments in the engineering of insect resistant plants. J. Cell. Biochem. Suppl. 11B:13.

van Doren, K., and Y. Gluzman. 1984. Efficient transformation of human fibroblasts by adenovirus–simian virus 40 recombinants. Mol. Cell. Biol. 4:1653–1656.

van Vloten-Doting, L., J.-F. Bol, and B. Cornelissen. 1985. Plant-virus-based vectors for gene transfer will be of limited use because of the high error frequency during viral RNA synthesis. Plant Mol. Biol. Rep. 4:323–326.

Watanabe, S., and H. M. Temin. 1983. Construction of a helper cell line for avian reticuloendotheliosis virus cloning vectors. Mol. Cell. Biol. 3:2241–2249.

Watrud, L. S., F. J. Perlak, M.-T. Tran, K. Kusano, E. J. Mayer, M. A. Miller-Wideman, M. G. Obukowicz, D. R. Nelson, J. P. Kreitinger, and R. J. Kaufman. 1985. Cloning of the *Bacillus thuringiensis* subsp. *kurstaki* delta-endotoxin gene into *Pseudomonas fluorescens*: molecular biology and ecology of an engineered microbial pesticide. In Engineered Organisms in the Environment: Scientific Issues (pp. 40–46), H. O. Halvorson, D. Pramer, and M. Rogul, eds. Washington, D.C.: American Society for Microbiology.

Wiktor, T. J., R. I. MacFarlan, K. J. Reagan, B. Dietzschold, P. J. Curtis, W. H. Wunner, M.-P. Kieny, R. Lathe, J.-P. Lecocq, M. Mackett, B. Moss, and H. Koprowski. 1984. Protection from rabies by a vaccinia virus recombinant containing the rabies virus glycoprotein gene. Proc. Natl. Acad. Sci. USA 81:7194–7198.

Williams, D. A., I. R. Lemischka, D. G. Nathan, and R. C. Mulligan. 1984. Introduction of new genetic material into pluripotent haematopoietic stem cells of the mouse. Nature 310:476–480.

Willis, R. C., D. J. Jolly, A. D. Miller, M. M. Plent, A. C. Esty, P. J. Anderson, H.-C. Chang, O. W. Jones, J. E. Seegmiller, and T. Friedmann. 1984. Partial phenotypic correction of human leschnyhan (hypoxanthine-guanine phosphoribosyltransferase-deficient) lymphoblasts with a transmissible retroviral vector. J. Biol. Chem. 259:7842–7849.

Yamada, M., J. A. Lewis, and T. Grodzicker. 1985. Overproduction of the protein product of a nonselected foreign gene carried by an adenovirus vector. Proc. Natl. Acad. Sci. USA 82:3567–3571.

Yamada, Y., Z. Q. Yang, and D. T. Tang. 1986. Plant regeneration from protoplast-derived callus of rice (*Oryza sativa* L.). Plant Cell Rep. 5:85–88.

Yelton, M. M., J. E. Hamer, and W. E. Timberlake. 1984. Transformation of *Aspergillus nidulans* by using a *trpC* plasmid. Proc. Natl. Acad. Sci. USA 81:1470–1474.

Zhu, Z., G. Li, L. He, and S. Chen. 1985. Novel gene transfer into the fertilized eggs of goldfish (*Carassius auratus* L 1758). J. Appl. Icthyol. 1:31–33.

Index

A